Springer Series in
MATERIALS SCIENCE 95

Springer Series in
MATERIALS SCIENCE

The Springer Series in Materials Science covers the complete spectrum of materials physics, including fundamental principles, physical properties, materials theory and design. Recognizing the increasing importance of materials science in future device technologies, the book titles in this series reflect the state-of-the-art in understanding and controlling the structure and properties of all important classes of materials.

78 **Macromolecular Nanostructured Materials**
Editors: N. Ueyama and A. Harada

79 **Magnetism and Structure in Functional Materials**
Editors: A. Planes, L. Mañosa, and A. Saxena

80 **Micro- and Macro-Properties of Solids**
Thermal, Mechanical and Dielectric Properties
By D.B. Sirdeshmukh, L. Sirdeshmukh, and K.G. Subhadra

81 **Metallopolymer Nanocomposites**
By A.D. Pomogailo and V.N. Kestelman

82 **Plastics for Corrosion Inhibition**
By V.A. Goldade, L.S. Pinchuk, A.V. Makarevich and V.N. Kestelman

83 **Spectroscopic Properties of Rare Earths in Optical Materials**
Editors: G. Liu and B. Jacquier

84 **Hartree-Fock-Slater Method for Materials Science**
The DV-X Alpha Method for Design and Characterization of Materials
Editors: H. Adachi, T. Mukoyama, and J. Kawai

85 **Lifetime Spectroscopy**
A Method of Defect Characterization in Silicon for Photovoltaic Applications
By S. Rein

86 **Wide-Gap Chalcopyrites**
Editors: S. Siebentritt and U. Rau

87 **Micro- and Nanostructured Glasses**
By D. Hülsenberg and A. Harnisch

88 **Introduction to Wave Scattering, Localization and Mesoscopic Phenomena**
By P. Sheng

89 **Magneto-Science**
Magnetic Field Effects on Materials: Fundamentals and Applications
Editors: M. Yamaguchi and Y. Tanimoto

90 **Internal Friction in Metallic Materials**
A Reference Book
By M.S. Blanter, I.S. Golovin, H. Neuhäuser, and H.-R. Sinning

91 **Time-dependent Mechanical Properties of Solid Bodies**
By W. Gräfe

92 **Solder Joint Technology**
Materials, Properties, and Reliability
By K.-N. Tu

93 **Materials for Tomorrow**
Theory, Experiments and Modelling
Editors: S. Gemming, M. Schreiber and J.-B. Suck

94 **Magnetic Nanostructures**
Editors: B. Aktas, L. Tagirov and F. Mikailov

95 **Nanocrystals**
Synthesis, Properties and Applications
By C.N.R. Rao, P.J. Thomas and G.U. Kulkarni

Volumes 30–77 are listed at the end of the book.

C.N.R. Rao P.J. Thomas G.U. Kulkarni

Nanocrystals: Synthesis, Properties and Applications

With 113 Figures, 6 in Color and 5 Tables

Springer

C.N.R. Rao, F.R.S.
P. J. Thomas
G.U. Kulkarni
Jawaharlal Nehru Center for Advanced Scientific Research
Jakkur P.O., Bangalore, 560 064 India
E-mail: cnrrao@jncasr.ac.in

Series Editors:

Professor Robert Hull
University of Virginia
Dept. of Materials Science and Engineering
Thornton Hall
Charlottesville, VA 22903-2442, USA

Professor R. M. Osgood, Jr.
Microelectronics Science Laboratory
Department of Electrical Engineering
Columbia University
Seeley W. Mudd Building
New York, NY 10027, USA

Professor Jürgen Parisi
Universität Oldenburg, Fachbereich Physik
Abt. Energie- und Halbleiterforschung
Carl-von-Ossietzky-Strasse 9–11
26129 Oldenburg, Germany

Professor Hans Warlimont
Institut für Festkörper-
und Werkstofforschung,
Helmholtzstrasse 20
01069 Dresden, Germany

ISSN 0933-033X
ISBN 978-3-642-08823-0
e-ISBN 978-3-540-68752-8

Springer is a part of Springer Science+Business Media.
springer.com

Softcover reprint of the hardcover 1st edition 2007

Cover concept: eStudio Calamar Steinen
Cover production: WMXdesign GmbH, Heidelberg

Preface

Nanoscience has emerged to become one of the most exciting areas of research today and has attracted the imagination of a large body of students, scientists, and engineers. The various kinds of nanomaterials that one normally deals with are the zero-dimensional nanocrystals, one-dimensional nanowires and nanotubes, and two-dimensional nanofilms and nanowalls. Of these, one of the earliest research investigations pertains to nanocrystals. It is truly remarkable that Michael Faraday made nanocrystals of gold and other metals in solution way back in 1857. Nanocrystals occupy a special place amongst nanomaterials because they have enabled a proper study of size-dependent properties. There have been several reviews, books, and conference proceedings dealing with nanomaterials in the last few years. In this monograph, we have attempted to give a well-rounded presentation of various aspects of nanocrystals. We first discuss some of the fundamentals and then make a detailed presentation of the synthetic methods. We examine the process of assembly of nanocrystals as well as their properties. Core–shell nanoparticles are treated as a separate chapter, just as the applications of the nanocrystals. We believe that this monograph should be useful to practicing scientists, research workers, teachers, and students all over the world. It could also form the basis of a course on the subject.

Bangalore
January 2007

C.N.R. Rao
P.J. Thomas
G.U. Kulkarni

Contents

1 **Basics of Nanocrystals** 1
1.1 Introduction 1
1.2 Properties of Nanocrystals 3
1.2.1 Geometric Structure 4
1.2.2 Magnetic Properties of Nanocrystals 9
1.2.3 Electronic Properties 12
1.2.4 Optical Properties 19
1.2.5 Other Properties 23

2 **Synthesis of Nanocrystals** 25
2.1 Physical Methods 25
2.1.1 Inert Gas Condensation 25
2.1.2 Arc Discharge 27
2.1.3 Ion Sputtering 27
2.1.4 Laser Ablation 28
2.1.5 Pyrolysis and Other Methods 28
2.1.6 Spray Pyrolysis 29
2.2 Chemical Methods 29
2.2.1 Metal Nanocrystals by Reduction 30
2.2.2 Solvothermal Synthesis 38
2.2.3 Photochemical Synthesis 42
2.2.4 Electrochemical Synthesis 42
2.2.5 Nanocrystals of Semiconductors and Other Materials by Arrested Precipitation 44
2.2.6 Thermolysis Routes 46
2.2.7 Sonochemical Routes 52
2.2.8 Micelles and Microemulsions 54
2.2.9 The Liquid–Liquid Interface 56
2.2.10 Biological Methods 56
2.2.11 Hybrid Methods 58
2.2.12 Solvated Metal Atom Dispersion (SMAD) 58
2.2.13 Post-synthetic Size-Selective Processing 59

2.3 Nanocrystals of Different Shapes 60
2.3.1 Shape-Controlled Synthesis of Metal Nanocrystals 60
2.3.2 Shape-Controlled Synthesis of Semiconductor and Oxide Nanocrystals 63
2.4 Doping and Charge Injection 66
2.5 Tailoring the Ligand Shell 67

3 Programmed Assemblies 69
3.1 One-Dimensional Arrangements 69
3.2 Rings and Associated Arrangements 72
3.3 Two-Dimensional Arrays 74
3.3.1 Metal Nanocrystals 75
3.3.2 Semiconductor and Oxide Nanocrystals 79
3.3.3 Other Two-Dimensional Arrangements 81
3.3.4 Mechanism of Organization 81
3.4 Three-Dimensional Superlattices 83
3.5 Superclusters 88
3.6 Colloidal Crystals 89

4 Properties of Nanocrystals 93
4.1 Melting Point and Heat Capacity 93
4.2 Electronic Properties 95
4.2.1 Catalysis and Reactivity 104
4.3 Optical Properties 106
4.4 Magnetic Properties 118

5 Core–Shell Nanocrystals 125
5.1 Synthesis and Properties 125
5.1.1 Semiconductor–Semiconductor 125
5.1.2 Metal–Metal 129
5.1.3 Metal–Oxide, Semiconductor–Oxide, and Oxide–Oxide 132
5.2 Assemblies of Core–Shell Nanocrystals 133

6 Applications 135
6.1 Introduction 135
6.2 Nanocrystals as Fluorescent Tags 135
6.3 Nanocrystal-Based Optical Detection and Related Devices 139
6.4 Biomedical Applications of Oxide Nanoparticles 141
6.5 Optical and Electro-Optical Devices 142
6.6 Dip-Pen Nanolithography with Nanocrystals 143
6.7 Nanoelectronics and Nanoscalar Electronic Devices 145
6.8 Nanocomputing 149

References 151

Index 175

1

Basics of Nanocrystals

1.1 Introduction

Nanoparticles constitute a major class of nanomaterials. Nanoparticles are zero-dimensional, possessing nanometric dimensions in all the three dimensions. The diameters of nanoparticles can vary anywhere between one and a few hundreds of nanometers. Small nanoparticles with diameters of a few nanometers are comparable to molecules. Accordingly, the electronic and atomic structures of such small nanoparticles have unusual features, markedly different from those of the bulk materials. Large nanoparticles (>20–50 nm), on the other hand, would have properties similar to those of the bulk [1]. The change in a material property as a function of size is shown schematically in Fig. 1.1. At small sizes, the properties vary irregularly and are specific to each size. At larger sizes, dependence on size is smooth and scaling laws can be derived to describe the variation in this regime. The size-dependent properties of nanoparticles include electronic, optical, magnetic, and chemical characteristics. Nanoparticles can be amorphous or crystalline. Being small in size, crystalline nanoparticles can be of single domain. Nanoparticles of metals, chalcogenides, nitrides, and oxides are often single crystalline. Crystalline nanoparticles are referred to as nanocrystals.

Nanoparticles are not new and their history can be traced back to the Roman period. Colloidal metals were used to dye glass articles and fabrics and as a therapeutic aid in the treatment of arthritis. The Purple of Cassius, formed on reacting stannic acid with chloroauric acid, was a popular purple dye in the olden days. It is actually made up of tin oxide and Au nanocrystals [2]. The Romans were adept at impregnating glass with metal particles to achieve dramatic color effects. The Lycurgus cup, a glass cup of 4th century AD, appears red in transmitted light and green in reflected light. This effect, which can be seen in the cup preserved in the British museum in London, is due to Au and Ag nanocrystals present in the walls of the cup. Maya blue, a blue dye employed by the Mayas around 7th century AD has been shown recently to consist of metal and oxide nanocrystals in addition to indigo and silica [3].

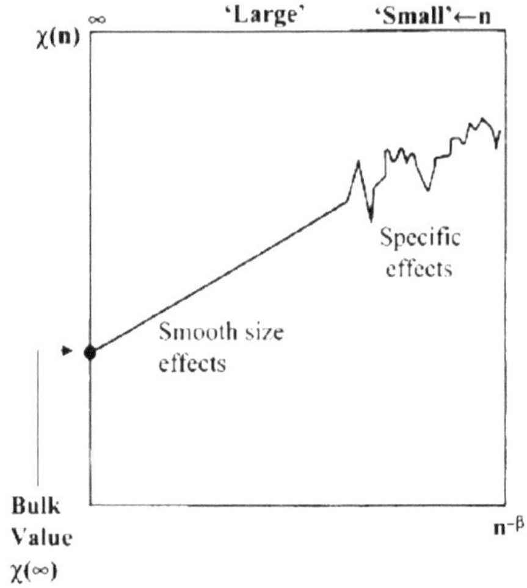

Fig. 1.1. The size dependence of a property $\chi(n)$ on the number of atoms (n) in a nanoparticle. The data are plotted against $n^{-\beta}$ where $\beta \geq 0$. Small nanoparticles reveal specific size effects, while larger particles are expected to exhibit a smooth size dependence, converging to the bulk value (reproduced with permission from [1])

Clearly, the ability to synthesize nanoparticles preceded the understanding of nanoscale phenomena. Systematic studies of nanoparticles began to appear as early as the seventeenth century. Antonio Neri, a Florentine glass maker and priest, describes the synthesis of colloidal gold in his 1612 treatise *L'Arte Vetraria*. John Kunckel, revised and translated Neri's work into german in 1689. Kunckel is often credited with the discovery that glass can be colored red by addition of gold.

Despite the early advances, studies of nanoscale particles did not gather momentum in later years. Thus, for most part of the 20th century, colloid science was the domain of a few specialized groups and did not receive sufficient importance. As early as 1857, Michael Faraday [4] carried out ground-breaking work on colloidal metals. He called them divided metals. Faraday established the very basis for the area, noting that colloidal metal sols were thermodynamically unstable, and that the individual particles must be stabilized kinetically against aggregation. Note that sols are dispersions of solids in liquids. Once the particles in a sol coagulate, the process cannot be reversed. Remarkably, Faraday also identified the essence of the nature of colloidal, nanoscale particles of metals. In the case of gold, he stated "gold is reduced in exceedingly fine particles which becoming diffused, produce a beautiful fluid ... the various preparations of gold whether ruby, green, violet, or blue ... consist of that substance in a metallic divided state." Einstein [5] related the Brownian motion executed by the nanoparticles to their diffusion coefficient. Mie and Gans [6–8] proposed a theoretical basis for the optical properties of the nanoscale particles, which continues to be used widely to this day. Frölich and Kubo proposed theories that predicted that the electronic structure of colloidal metals would differ from bulk.

The neglect of colloid science prompted Ostwald [9] to title his 1915 book on colloids as "The world of neglected dimensions." This period also witnessed advances in methods to make colloidal gold. Bredig [10] prepared Au sols by striking an arc between Au electrodes immersed in dilute alkali. Donau [11] suggested that passing CO through a solution of chloroauric acid provided a gold sol. Zsigmondy [12] discovered the seeding method and was familiar with the use of formaldehyde in mild alkali to produce Au sols from salts. The 1925 award of the Nobel prize to Zsigmondy partly for his work on gold colloids did not seem to have enthused the scientific community to pursue this area of research. In the last few years, however, there has been a great upsurge in the use of colloid chemical methods to generate nanoparticles of various materials. This is because of the excitement caused by the science of nanomaterials initiated by the now famous lecture of Feynman [13].

Explosion of research in nanocrystals has been so dramatic that very few of the modern practitioners seem to be aware of the glorious past of colloid science. The progress has been facilitated in part by the advances in instrumentation that have helped in fully characterizing nanomaterials. Today, it is possible to prepare and study nanocrystals of metals, semiconductors and other substances by various means. Advances in both experimental and theoretical methods have led to an understanding of the properties of nanocrystals.

1.2 Properties of Nanocrystals

Nanocrystals of materials are generally obtainable as sols. Sols containing nanocrystals behave like the classical colloids. For example, the stability of a dispersion depends on the ionic strength of the medium. Nanocrystalline sols possess exceptional optical clarity. A key factor that lends stability to nanocrystal sols is the presence of a ligand shell, a layer of molecular species adsorbed on the surface of the particles. Without the ligand shell, the particles tend to aggregate to form bulk species that flocculate or settle down in the medium. Depending on the dispersion medium, the ligands lend stability to particles in two different ways. Thus, in an aqueous medium, coulomb interactions between charged ligand species provide a repulsive force to counter the attractive van der Waals force between the tiny grains, by forming an electrical double layer. In an organic medium, the loss of conformational freedom of the ligands and the apparent increase in solute concentration provide the necessary repulsive force. We illustrate this schematically in Fig. 1.2. Nanocrystals dispersed in liquids are either charge-stabilized or sterically stabilized.

Nanoparticles devoid of ligands are generally studied in vacuum. Such particles deposited on a substrate are readily examined by photoelectron spectroscopy and other techniques. Beams of uniformly sized clusters traversing

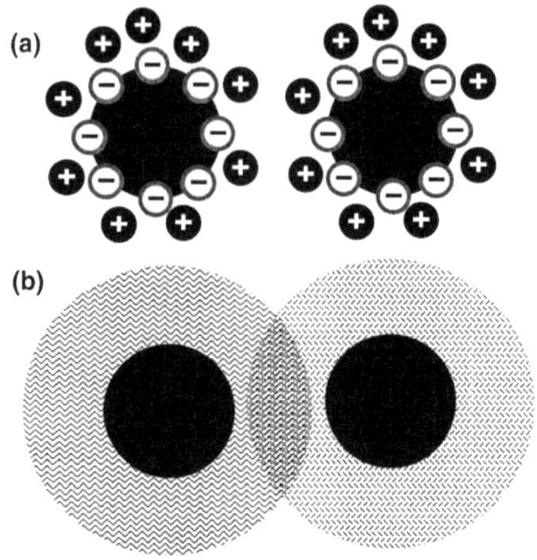

Fig. 1.2. Schematic illustration of the factors lending stability to a colloidal dispersion: (**a**) an electric double layer and (**b**) loss of conformational freedom of chain-like ligands

a vacuum chamber (cluster beams) with some fixed velocity provide opportunities for studies of the intrinsic physical properties of nanoparticulate matter.

1.2.1 Geometric Structure

The dimensions of nanocrystals are so close to atomic dimensions that an unusually high fraction of the total atoms would be present on their surfaces. For example, a particle consisting of 13 atoms, would have 12 atoms on the surface, regardless of the packing scheme followed. Such a particle has a surface more populated than the bulk. It is possible to estimate the fraction of atoms on the surface of the particle (P_s, percentage) using the simple relation,

$$P_s = 4N^{-1/3} \times 100, \qquad (1.1)$$

where N is the total number of atoms in the particle [14]. The variation of the surface fraction of atoms with the number of atoms is shown in Fig. 1.3. We see that the fraction of surface atoms becomes less than 1% only when the total number of atoms is of the order of 10^7, which for a typical metal would correspond to a particle diameter of 150 nm.

Nanoparticles are generally assumed to be spherical. However, an interesting interplay exists between the morphology and the packing arrangement, specially in small nanocrystals. If one were to assume that the nanocrystals strictly follow the bulk crystalline order, the most stable structure is arrived at by simply constraining the number of surface atoms. It is reasonable to assume that the overall polyhedral shape has some of the symmetry elements of the constituent lattice. Polyhedra such as the tetrahedron, the octahedron,

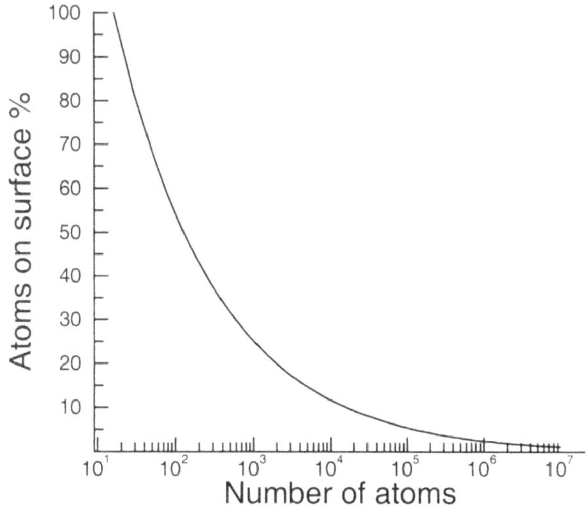

Fig. 1.3. Plot of the number of atoms vs. the percentage of atoms located on the surface of a particle. The calculation of the percentage of atoms is made on the basis of (1.1) and is valid for metal particles

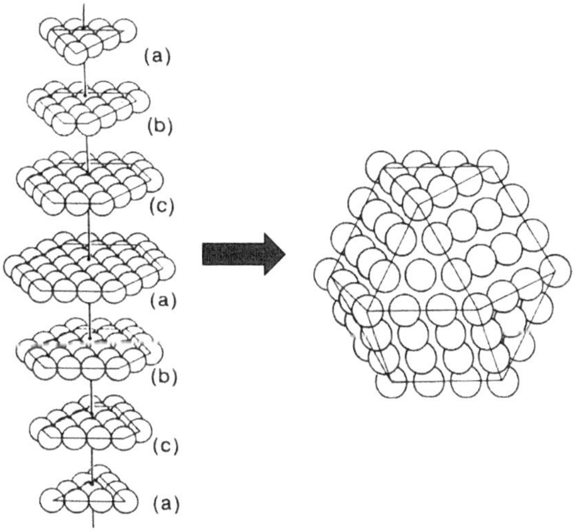

Fig. 1.4. Schematic illustration of how a cuboctahedral 147 atom-cluster, composed of seven close-packed layers can be made out of a stacking sequence reminiscent of a fcc lattice (reproduced with permission from [16])

and the cuboctahedron can be constructed following the packing scheme of a fcc lattice [15, 16]. Figure 1.4 shows how a cuboctahedral cluster of 146 constituent atoms follows from a fcc type *abcabc* layer stacking. In contrast to the above, small clusters frequently adopt *non-close packed* icosahedral

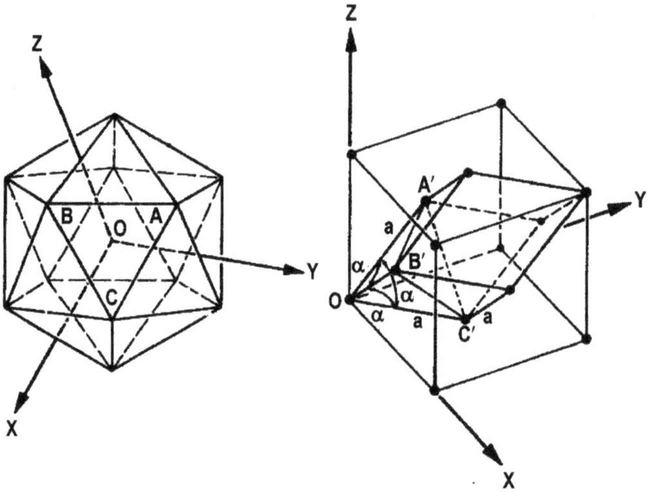

Fig. 1.5. The regular icosahedron is made up of twenty irregular tetrahedra like OABX. The rhombhohedral cell in a fcc lattice (OA′B′C′) has $\alpha = 60°$. When α is distorted to 63.43°, OA′B′C′ and OABC become similar. Small nanocrystals distort in a similar manner from regular fcc lattice to adopt the icosahedral shape

or dodecahedral shapes. The clusters adopting such schemes suffer a loss in packing efficiency. The icosahedron has a fivefold symmetry, inconsistent with the packing requirements of a regular crystalline lattice with long-range order. While employing close packing schemes, a stacking fault becomes necessary to arrive at an icosahedral arrangement. Such a scheme is outlined in Fig. 1.5.

The icosahedron which has twenty triangular faces and twelve vertices consists of a fcc-like close packing. Each of the twenty triangular faces of an icosahedron can be considered as a base of a tetrahedron, whose apex is at the inversion center (see Fig. 1.5a). A tetrahedron OA′B′C′ in Fig. 1.5b joining three face-centered atoms and an atom at the base of a fcc unit cell has the angle, $\alpha = 60°$. These angles can be distorted to 63.43°, to obtain the tetrahedron (OABC) that forms the building block of an icosahedron. Such a distortion results in the lowering of the packing fraction from 0.74080 to 0.68818. Several theoretical investigations have sought to explain the unusual stability of icosahedral clusters [17–19]. Allpress and Sanders [20,21], based on potential energy calculations, showed that the binding energy per atom is lower than that in a corresponding octahedron containing the same number of atoms. Molecular dynamics simulations have shown that Al clusters with nuclearities upto 147 atoms exhibit distorted icosahedral structures while Al_{147} has a cuboctahedral shape [22]. More rigourous theories (ab initio, density functional) broadly support this contention. A decahedral shape can be thought of as being made up of four edge-sharing tetrahedra, followed by some relaxation and the consequent loss of packing fraction. Ino [23, 24] has suggested the use of the term "multiply twinned particle" to denote a decahedral particle, and such particles obtained by the twining of tetrahedra.

The properties of nanocrystals are also influenced by the formation of geometric shells which occur at definite nuclearities [25, 26]. Such nuclearities, called magic nuclearities endow a special stability to nanocrystals as can be demonstrated on the basis of purely geometric arguments. A new shell of a particle emerges when the coordination sphere of an inner central atom or shell (forming the previous shell) is completely satisfied. The number of atoms or spheres required to complete successive coordination shells is a problem that mathematicians, starting with Kepler, have grappled with for a long time [27, 28]. The "kissing" problem, as it is known in the mathematical world, was the subject of a famous argument between Newton and Gregory at Cambridge. In retrospect, Newton, who held that 12 atoms are required to complete the second shell was indeed correct. An idea of the mathematical effort involved can be gauged from the fact that the proof of Newton's argument was provided only in 2002. It is quite apparent that the ultimate shape of the emerging crystallite should play a role in determining the number of atoms that go into forming complete shells. The magic nuclearities would then yield information on the morphology of the cluster. By a strange coincidence, the number of atoms required to form complete shells in the two most common shapes (icosahedron and cuboctahderon) is the same.

The number of atoms, N, required to form a cluster with L geometric shells is given by

$$N = \frac{(10L^3 + 15L^2 + 11L + 3)}{3}. \tag{1.2}$$

This represents the solution for the "kissing" problem in three dimensions and is valid for icosahedral and cuboctahedral morphologies. For other shapes, the reader may refer to a paper from the group of Martin [29]. Particles possessing the above number of atoms are said to be in a closed-shell configuration. The number of atoms required to fill up coordination shell completely, n_L of a particular shell, is given by

$$n_L = (10n_{L-1}^2 + 2). \tag{1.3}$$

where n_0=1. Thus, 12 atoms are required to complete the first shell, 42 to complete the second shell etc. A schematic illustration of the observed magic nuclearity clusters is provided in Fig. 1.6. The notion of the closed-shell configuration can be extended to larger dimensions as well. Closed-shell configurations lend stability to giant clusters made of clusters and even to a cluster of giant clusters.

Determination of the structures of nanocrystals should ideally follow from X-ray diffraction, but small particles do not diffract well owing to their limited size. The peaks in the diffraction pattern are less intense and are broad. Structural studies are therefore based on high resolution transmission electron microscopy (HRTEM), extended X-ray absorbtion fine structure (EXAFS),

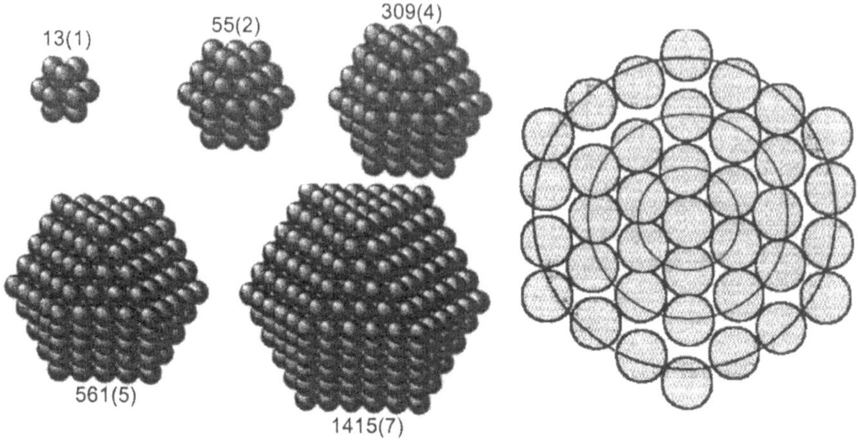

Fig. 1.6. Nanocrystals in closed-shell configurations with magic number of atoms. The numbers beside correspond to the nuclearity (N) and the number of shells (L). The figure on the left is a cross-sectional view showing five coordination shells in a 561 atom cluster

scanning tunneling microscopy (STM) and atomic force microscopy (AFM). X-ray diffraction patterns provide estimates of the diameters (D) of nanocrystals from the width of the diffraction profiles, by the use of the Scherrer formula [30]

$$D = \frac{0.9\lambda}{\beta \cos\theta}. \tag{1.4}$$

Here, β is the full-width at half-maximum of the broadened X-ray peak corrected for the instrumental width,

$$\beta = \beta^2_{\text{observed}} - \beta^2_{\text{instrumental}}. \tag{1.5}$$

Estimates based on the Scherrer relation are used routinely. It is desirable to carry out a Reitveld analysis of the broad profiles of nanoparticles to obtain estimates of D.

HRTEM with its ability to image atomic distributions in real space, is a popular and powerful method. The icosahedral structure of nanocrystals is directly observed by HRTEM and evidence for twinning (required to transform a crystalline arrangement to an icosahedron) is also obtained by this means (see Fig. 1.7) [15, 31]. The images are often compared with the simulated ones [32, 33]. High resolution imaging provides compelling evidence for the presence of multiply twinned crystallites specially in the case of Au and Ag nanoparticles [34]. Characterization by electron microscopy also has certain problems. For example, the ligands are stripped from the clusters under the electron beam; the beam could also induce phase transitions and other dynamic events like quasi-melting and lattice reconstruction [35]. The fact

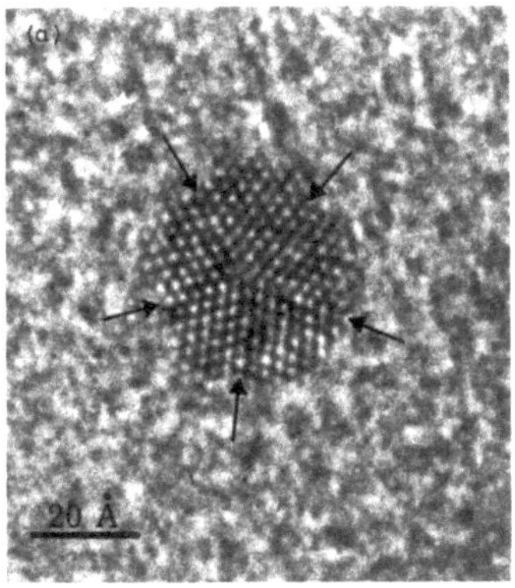

Fig. 1.7. A high resolution TEM image showing the icosahedral shape and the five fold symmetry axis of a Ag nanoparticle (reproduced with permission from [31])

that ligands desorb from clusters has made it impossible to follow the influence of the ligand shell on cluster packing.

STM, with its ability to resolve atoms, provides exciting opportunities to study the size and morphology of individual nanoparticles. In the case of ligated nanocrystals, the diameters obtained by STM include the thickness of the ligand shell [36]. Ultra high vacuum STM facilitates in situ studies of clusters deposited on a substrate. Furthermore, it is possible to manipulate individual nanoscale particles using STM. However, it is not possible to probe the internal structure of a nanocrystal, especially if it is covered with a ligand shell. AFM supplements STM and provides softer ways of imaging nanocrystals. EXAFS has advantages over the other techniques in providing an ensemble average, and is complimentary to HRTEM [37].

1.2.2 Magnetic Properties of Nanocrystals

Isolated atoms of most elements possess magnetic moments that can be arrived at on the basis of Hund's rules. In the bulk, however, only a few solids are magnetic. Nanoscale particles provide opportunities to study the evolution of magnetic properties from the atomic scale to the bulk. Even before the explosion of interest in nanoscience, magnetic properties of the so-called fine particles had been examined. In fact, size effects were perhaps first noticed in magnetic measurements on particles with diameters in the 10–100 nm range [38].

In order to understand size-dependent magnetic properties, it is instructive to follow the changes in a magnetic substance as the particle size is decreased from a few microns to a few nanometers. In a ferromagnetic substance, the T_c decreases with decrease in size. This is true of all transition temperatures associated with long-range order. For example, ferroelectric transition temperatures also decrease with particle size. With the decrease in the diameter,

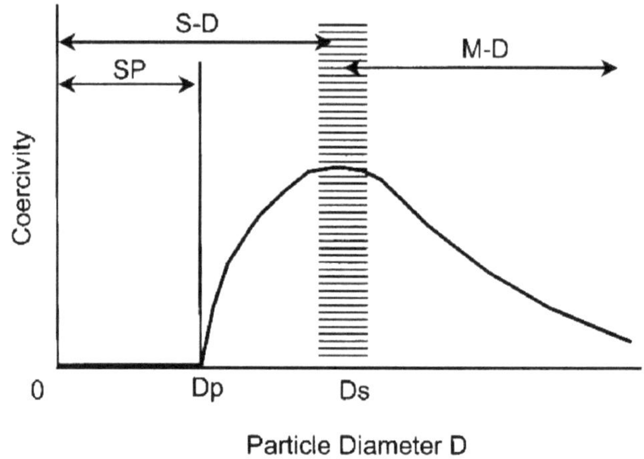

Fig. 1.8. Schematic illustration of the change in the coercivity of a ferromagnetic particle with the diameter. SP denotes the superparamagnetic regime, S-D the single-domain regime and M-D the multi-domain regime

the coercivity (H_c), increases initially till a particular diameter, D_s, and thereafter decreases as shown in Fig. 1.8. The critical diameter, D_s, marks a region wherein the particle changes from being a multi-domain particle to a single-domain particle. The value of D_s is normally a few tens of nanometers. The single-domain nature of the nanoparticle is its single most attractive magnetic property. Below D_s, H_c tends to decrease due to thermal effects and follows a relation of the form:

$$H_c = g - \frac{h}{D^{3/2}}, \tag{1.6}$$

where g and h are constants [39]. Below the critical diameter D_p, coercivity becomes zero as the thermal energy would be sufficient to randomize the magnetic moments in the particle. Nanocrystals below the diameter D_p ($\sim$10 nm), exhibit such a behavior and are said to be superparamagnetic. Superparamagnetic particles do not possess long-range magnetic order, but show characteristic magnetic properties at low temperatures [40]. Particles delineated on the basis of the above critical regimes, follow different paths to the final magnetized state and the magnetization reversal mechanism. These paths and mechanisms are rather complicated [39].

Size-dependent changes in magnetic anisotropy are another aspect of interest. A magnetic material exhibits strong and often complex anisotropic behavior when subjected to magnetization. The magnetic anisotropic energy (E), defined as the energy difference involved in changing the magnetization direction from a low-energy direction or easy axis to a high energy direction or hard axis, is an important technological parameter. Materials with high E and low E find numerous applications. The simplest form of anisotropy is the uniaxial anisotropy, where E is only dependent on the angle that the magnetic field vector makes with the easy axis of the sample. Square hysteresis loops are obtained when the direction of magnetization is parallel to the easy axis and a straight line response is obtained when the field direction is perpendicular to the easy axis direction [41, 42]. Materials are also known to exhibit hexagonal and cubic anisotropy. The nature of magnetic anisotropy can be arrived at by studying hysteresis loops obtained along various directions. E exhibits dramatic changes with the change in size and shape. Nanoparticles generally possess higher E, but may also show a change in the basic nature of anisotropy.

Superparamagnetic behavior is caused by thermal flipping of the anisotropic barrier to magnetization reversal. Below a certain temperature, called the blocking temperature, temperature-induced flipping or relaxation can be arrested and the nanocrystals acquire a finite coercivity. In Fig. 1.9, magnetic measurements indicating size-dependent changes in the blocking temperature of $CoFe_2O_4$ nanoparticles are shown. The superparamagnetic behavior was first modeled by Neel in the 1950s [43]. In the case of nanoparticles with uniaxial anisotropy, Neel's theory suggests that the temperature induced relaxation varies exponentially with temperature and scales with the sample volume. Besides the loss of coercivity, another characteristic feature

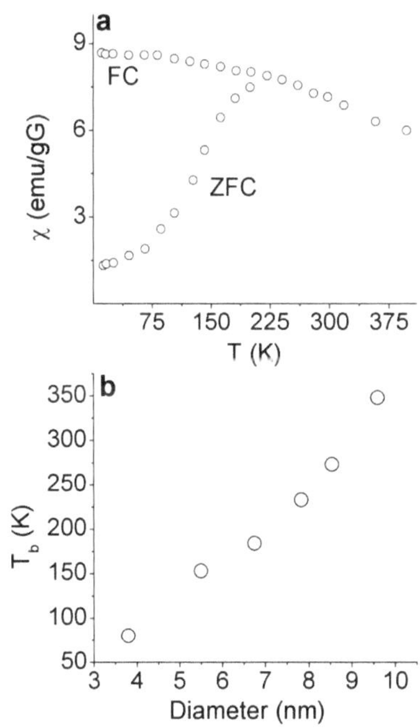

Fig. 1.9. (**a**) Magnetic susceptibility vs. temperature for $CoFe_2O_4$ nanoparticles under field cooled and zero-field cooled conditions. The applied field is 2,000 G. (**b**) shows the variation of blocking temperature (T_b) with diameter of nanoparticles (plot produced with data from [40])

of superparamagnetic behavior is the scaling of magnetization with temperature. A plot of the magnetization (M) and the ratio of the magnetic field and the temperature (H/T) produces a universal curve for all temperatures above T_b. Experimentally, superparamagnetism may be probed by techniques such as neutron scattering, Mössbauer spectroscopy and magnetization measurements. Since superparamagnetic behavior is related to the relaxation rate, it is sensitive to the characteristic time scale of measurement. In the experimental techniques indicated above, the time scales of measurements vary over a wide range (10^{-14}–10^{-12} for neutron scattering to around 10 s for magnetization measurement) and different measurements, therefore, yield different blocking temperatures. Though the scaling with volume suggested by the Neel theory is largely followed in the uniaxial cases, the actual blocking behavior of magnetic nanocrystals, especially those of the oxides are quite complicated. Effects due to the ligand shell and lattice defects are generally considered to be responsible for the observed deviations.

Many metal oxide nanoparticles are known to show evidence for the presence of ferromagnetic interactions at low temperatures. This is specially true of nanoparticles of antiferromagnetic oxides such as MnO, CoO, and NiO [44,45]. What one normally observes is a divergence in the zero-field cooled and field cooled magnetization data at low temperatures. The materials show magnetic hysteresis below a blocking temperature typical of superparamagnetic materials. The blocking temperature generally increases with the particle size. In Fig. 1.10, we show the magnetization behavior of NiO particles of 3 and 7 nm diameters [45].

While reasonable progress has been made in understanding the magnetic properties of isolated particles, ensembles of particles represent a relatively poorly researched area [46, 47]. A typical ensemble consists of particles with distributions in size, shape and easy anisotropy direction. Further, interparticle interactions play a role in determining the magnetic response of the ensemble. Theoretical investigations predict interesting phenomena in such assemblies, including changes in the mechanism of magnetization, depending on interparticle interactions. Experimentally, controlled interactions are brought about by varying parameters such as the volumetric packing density. Experiments have been carried out on ensembles of nanoparticles obtained in various ways such as freezing a sol containing a known fraction of magnetic nanoparticles or dilution in a polymer matrix [41, 46–48]. Seminal advances have been made in obtaining such ensembles by self-assembly based techniques. Self-assembled nanocrystalline ensembles posses several advantages over those obtained by other means.

1.2.3 Electronic Properties

Bulk metals possess a partially filled electronic band and their ability to conduct electrons is due to the availability of a continuum of energy levels above E_F, the fermi level. These levels can easily be populated by applying an

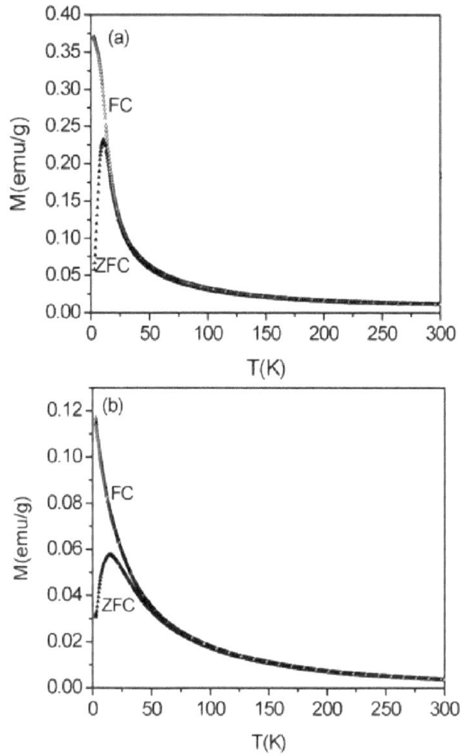

Fig. 1.10. The temperature dependence of dc magnetization of (**a**) 3 nm and (**b**) 7 nm NiO nanoparticles under zero-field cooled and field cooled conditions (H = 100 Oe)(reproduced with permission from [45])

electric field and the electrons now behave as delocalized Bloch waves ($\lambda \sim$ 5–10 Å) [14]. When a metal is divided finely, the continuum of the electronic states breaks down and the sample ultimately becomes insulating. The discreteness of energy levels do not physically manifest themselves as long as the gap is less than $k_b T$, the thermal energy at temperature T. The discreteness of the levels can be measured in terms of average spacing between the successive quantum levels, δ, given by

$$\delta = \frac{4E_F}{3n_e}, \tag{1.7}$$

and is known as the Kubo gap [49]. Here, n_e is the number of valence electrons in the cluster (a contribution of one valence electron per constituent atom is assumed). In the case of semiconductors, a reduction in the size of the system causes the energy levels at the band edge to become discrete, with interlevel spacings similar to metals. This effectively increases the bandgap of the semiconductor [50]. The issues at hand are sketched schematically in

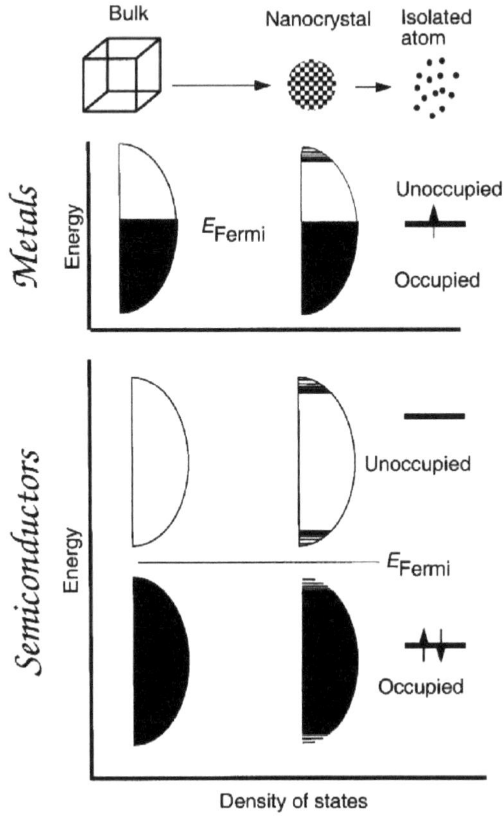

Fig. 1.11. Schematic illustration of the changes in the electronic structure accompanying a reduction in size, in metals and semiconductors

Fig. 1.11. Additional complications are introduced by the strongly directional covalent bonds present in a semiconductor. Accompanying the appearance of the discrete levels are other consequences such as the change from metallic to van der Waals type of bonding, lowering of the melting point, odd–even effects, and perhaps most significantly, a metal–insulator transition. The whole gamut of such changes are generally called as quantum size effects [51].

Another consideration required to describe the electronic structure of metal clusters is the emergence of electronic shells. The closure of electronic shells, similar to the closure of geometric shells, bequeath special stability to specific nuclearities and is manifest best in the variation of ionization energies with cluster nuclearity [29,52]. The total number of electrons required to arrive at successive closed-shell configuration is 2, 10, 28, 60.... The closed shell configuration is based on a hydrogen atom-like potential and hence the familiar 2, 8, 18, 32 pattern for the outer electron configuration to fill the electronic shells in the clusters. Other theoretical models predict different sets of closed-shell configurations. A unified picture of both geometric and electronic shells can be obtained by considering a characteristic length in closed-shell configurations [16,29]. A unit increase of the characteristic length scale results in the closure of a shell. For geometric clusters, the characteristic length is the interatomic distance, while the characteristic length for electronic shells is related to the electron wavelength in the highest occupied level. In the case of alkali metal clusters, this wavelength is around twice the interatomic distance.

Experimental evidence for electronic shells is found in the plot of cluster abundance vs. nuclearity and in the variation of the ionization energies of clusters (see Fig. 1.12). Electronic shell effects dominate the properties of alkali metal clusters. They are also broadly applicable to p-block metals. The properties of transition and nobel metal nanoparticles, however, are influenced more by the formation of geometric shells. In fact, a transition from shells of electrons to shells of atoms is seen in the case of Al [29, 53]. It appears that the abundance of available oxidation states and the directional nature of the d- and f-orbitals (and to a limited extent, of the p-orbital) play a role in determining the shell that governs the property of a particular cluster.

A host of physical techniques has been used to follow phenomena such as the closure of the bandgap and the emergence of the metallic state from the cluster regime. Photoelectron spectroscopy methods are among the most popular methods to study nanoparticles. Core level X-ray photoelectron spectroscopy (XPS) provides information on the oxidation state of the atoms. Changes in the electronic structure are manifested as binding energy shifts and through the broadening of photoemission bands. XPS is ideally suited to study changes in the electronic structure of clusters, accompanying the

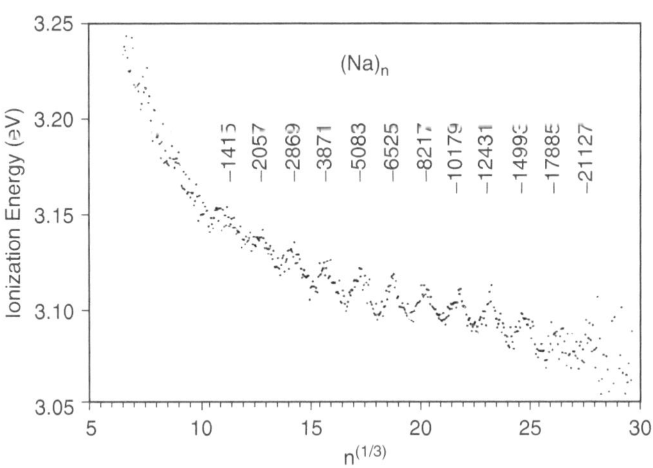

Fig. 1.12. Ionization energies of Na clusters vs. the third root of cluster nuclearity. The oscillations occur at closed shell nuclearities. The nuclearities are indicated on top (reproduced with permission from [16])

adsorption of gases. The density of states around the fermi level can be probed by means of ultraviolet photoelectron spectroscopy (UPS), while Bremsstrahlung isochromat spectroscopy (BIS) provides information on the unoccupied levels. Direct information on the gap states in nanocrystals of metals and semiconductors is obtained by scanning tunneling spectroscopy (STS). This technique provides the desired sensitivity and spatial resolution making it possible to carry out tunneling spectroscopic measurements on individual particles. The various techniques and the region of the band structure probed by them are schematically illustrated in Fig. 1.13.

Nanoscale particles of metals and semiconductors have been subject to numerous theoretical investigations [54]. Theories have evolved in sophistication and rigor since the beginnings made by Kubo and Frolich. The electronic and geometric structure of nobel metal clusters containing up to a few hundred atoms are studied using ab initio calculations. A few surprises have resulted from such investigations. For example, detailed calculations on anionic coinage metal clusters with seven atoms have indicated that while Cu_7^- and Ag_7^-

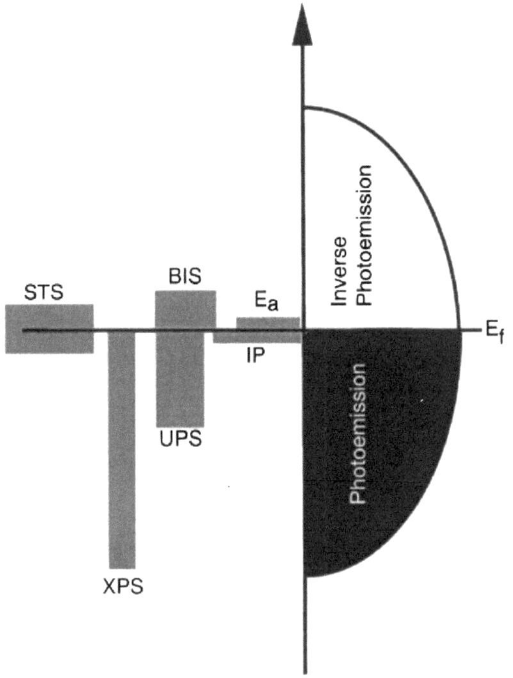

Fig. 1.13. A schematic illustration of the spectroscopic techniques and the portion of the band structure that they probe. The techniques illustrated are ionization potential (IP) measurements, electron affinity measurements (E_a), Bremsstrahlung isochromat spectroscopy (BIS), Ultraviolet photoelectron spectroscopy (UPS), X-ray photoelectron spectroscopy (XPS), Scanning tunneling spectroscopy (STS)

have three dimensional structures, Au_7^- is planar [55]. Planar structures are energetically favored for other Au nuclearities as well [56]. Ab initio molecular dynamics simulations have been carried out on clusters containing up to a few tens of atoms. For example, Landman and coworkers [57, 58] have carried out extensive computer simulations, on the structure and dynamics of alkanethiol-capped Au nanocrystals of different sizes. Numerous methods have been devised to study the properties of larger clusters often borrowing from theories used to model the bulk. Self-consistent Jellium model and local density approximation (a DFT based method) are popular for the study of metal clusters. The latter is also useful to investigate semiconductor nanocrystals. Tight-binding approximation based methods are used to study tetrahedrally coordinated semiconductors. Currently, experiments and theories go hand in hand in the study of small particles (~1 nm) and as the methods develop, it is not difficult to visualize a time in the near future, where the same can be said for investigations on larger nanocrystals.

In a bulk metal, the energies required to add or remove an electron are equivalent and are called the work function (W). In contrast, in the case of a molecule, the corresponding energies, electron affinity (E_a) and ionization potential (IP) are nonequivalent [59]. The two energies differ to a smaller extent in a nanoparticle and are size-dependent. E_a and IP are given by

$$\mathrm{IP} = W + \frac{\alpha e^2}{R} \tag{1.8}$$

$$E_a = W + \frac{\beta e^2}{R} \tag{1.9}$$

with the constraint,

$$\alpha + \beta = 1 \tag{1.10}$$

Here, R is the radius of the particle and α and β are constants. The physical significance and the values of α and β have been a subject of much debate. It appears that the value $\alpha = 3/8$ is appropriate for alkali metal clusters while $\alpha = 1/2$ agrees with the experimental results on nobel metal particles [52,60]. The difference in the two energies, E_a and IP, is the charging energy, U, given by,

$$U = \mathrm{IP} - E_a = \frac{(\alpha - \beta)e^2}{R}. \tag{1.11}$$

Typical values of U are of the order of a few hundred meV. It is to be noted that U is manifest as a Coulombic gap and is different from the electronic energy gap. It is possible to define differences between successive IPs and E_as and arrive at various states called the Coulombic states, defined by successive Us. The Coulombic states are presumed to be similar for both semiconductor and metallic nanocrystals unlike the electronic states [61]. A quantity closely related to U that aids in a better understanding is the capacitance (C) of a particle, which is given by

$$C = \frac{U}{2e}. \tag{1.12}$$

The classical expression for the capacitance of a metal sphere embedded in a dielectric with a dielectric constant, $\epsilon_{\rm m}$ is

$$C = 4\pi\epsilon\epsilon_{\rm m}R. \tag{1.13}$$

Equations 1.12 and 1.13, agree with each other and yield capacitances of the order of 10^{-18} F (or 1 aF) for nanoparticles. In this regime of finite charging energies and low capacitances, the charging of a capacitor is no longer continuous, but is discrete. For a current to flow through a nanocrystal, an external voltage $V_{\rm ext}$ greater than $\frac{e}{2C}$ is required. This phenomena of current exclusion across zero bias is called Coulomb blockade. When the Coulomb blockade barrier is broken applying sufficient voltage, electrons tunnel into the nanocrystal and tunnel out almost immediately. The electrons may reside long enough to provide a voltage feedback preventing an additional electron from tunneling in simultaneously. A continuous one electron current, I given by

$$I = \frac{e}{2RC} \tag{1.14}$$

flows through the circuit. To place an additional electron on the nanocrystal, a full e/C increase in voltage is required. Thus, steps called a Coulomb staircase becomes visible in the IV spectra of nanocrystals. A schematic illustration of a Coulomb staircase observed in the IV spectra is given in Fig. 1.14. While Coulomb blockade is ordinarily observed, the observation of a Coulomb staircase, requires a tuning of the circuit characteristics. The most significant impact of this discovery is in the realization of single electron

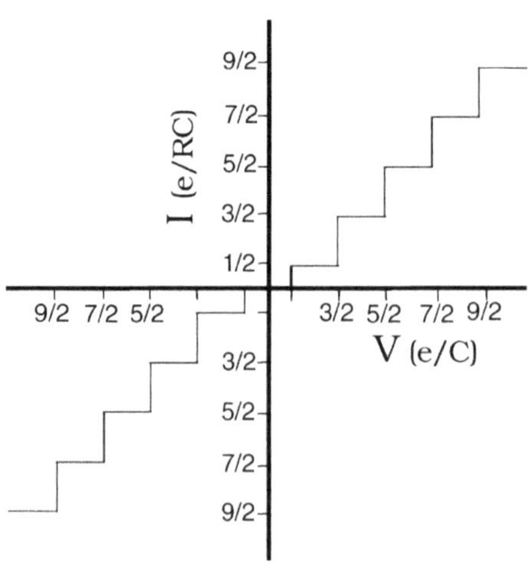

Fig. 1.14. Schematic illustration of the Coulomb staircase behavior seen in the I-V spectra

devices at room temperature [62–65]. Exploratory single electron devices, including single electron transistors, operable at room temperature have been realized by exploiting the low capacitance of nanocrystals [62–64]. Single electron devices use the smallest possible charge to store information and hence are among the most efficient of known devices. It is indeed fascinating that the measurement of IP and E_a has led to such possibilities.

1.2.4 Optical Properties

The electronic absorption spectra of nanocrystals of metals is dominated by the surface plasmon band which arises due to the collective coherent excitation of the free electrons within the conduction band [66–69]. A schematic illustration of the electric field component of an incoming light wave inducing a polarization of the free or itinerant electrons is shown in Fig. 1.15. It corresponds to the dipolar excitation mode which is the most relevant for particles whose diameters are much less than the wavelength of light. However, higher order excitations are possible and come into play for nanocrystals with diameters in the range of tens of nanometers.

As mentioned earlier, the theory of optical absorption of small particles was proposed by Mie in 1908 [6]. Mie's electrodynamic solution to the problem of light interacting with particles involved solving Maxwell's equations with appropriate boundary conditions and leads to a series of multipole oscillations for the extinction (C_{ext}) and scattering (C_{sca}) cross-sections of the nanoparticles. Thus

$$C_{\mathrm{ext}} = \frac{2\pi}{k^2} \sum_{n=1}^{\infty} (2n+1) Re(a_n + b_n) , \tag{1.15}$$

$$C_{\mathrm{sca}} = \frac{2\pi}{|k|^2} \sum_{n-1}^{\infty} (2n+1) Re(|a_n|^2 + |b_n|^2) , \tag{1.16}$$

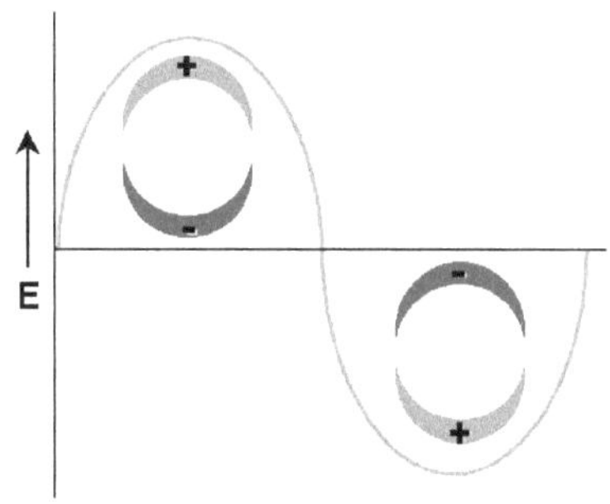

Fig. 1.15. A schematic illustration of the excitation of the dipole surface plasmon oscillation by the electric field component of the light wave. The dipolar oscillation of the electrons has the same frequency as that of the incoming light wave

where $\epsilon_{\rm m}$ is the dielectric constant of the surrounding medium, $k = \frac{2\pi\sqrt{\epsilon_{\rm m}}}{\lambda}$, a_n and b_n are Ricatti–Bessel functions that depend on the wavelength and the nanoparticle radius (R). For small particles, the scattering term and the higher order extinction terms are negligible. The extinction coefficient is, therefore, given by,

$$C_{\rm ext} = \frac{24\pi^2 R^3 \epsilon_m^{3/2}}{\lambda} \frac{\epsilon_2}{(\epsilon_1 + 2\epsilon_m)^2 + {\epsilon_2}^2}, \tag{1.17}$$

where ϵ_1 and ϵ_2 are the real and imaginary part of the frequency-dependent dielectric constant, ϵ, of the substance. The dielectric constants can either be obtained from the Drude free electron model or from experiments [70]. The Drude model relates the dielectric constant to the bulk plasmon frequency, $\omega_{\rm p}$, and the dampening frequency, $\omega_{\rm d}$, which are given by

$$\omega_{\rm p} = \frac{N_{\rm e} e^2}{m_{\rm e}^* \epsilon}, \tag{1.18}$$

$$\omega_{\rm d} = \frac{V_{\rm f}}{L}, \tag{1.19}$$

where $N_{\rm e}$ is the concentration of the free electrons, $m_{\rm e}^*$ the effective mass of the electron, $V_{\rm f}$ the velocity of the electrons at the fermi level and L, the mean free path. Small particles with diameters comparable to the mean free path exhibit a diameter dependent L, given by,

$$\frac{1}{L_{\rm effective}} = \frac{1}{2R} + \frac{1}{L}. \tag{1.20}$$

The origin of color lies in the denominator of (1.17), which predicts an absorbance maximum for a sol, when

$$\epsilon_1 = -2\epsilon_m. \tag{1.21}$$

Thus, metal nanocrystals of various sizes exhibit characteristic colors depending on their diameters and the dielectric constant of the surrounding medium. Typical size-dependent changes in the optical spectra are shown in Fig. 1.16. Mathematical methods have been developed to accurately compute the higher order terms in Mie's theory [71]. Several off-the-shelf programs are available to apply Mie's theory and its extensions to a variety of nanoparticles [72]. One such code called "BHCOAT," presented in the book by Bohren and Huffman [73] is very popular.

The somewhat complicated nature of surface plasmon excitation and its dependence on factors such as the distribution in particle diameters have thwarted attempts to derive fundamental physical quantities and to obtain quantitative information on the electronic structure of fine particles based on plasmon resonance spectroscopy. Initial studies such as those due to Kreker and Kreibig [74, 75] have been instrumental in verifying the validity of Mie's

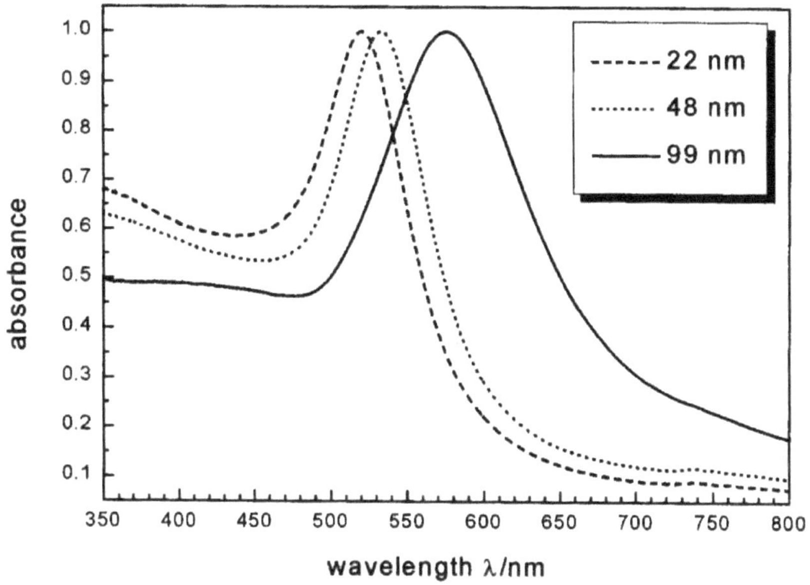

Fig. 1.16. Optical absorption spectra of gold nanoparticles with diameters of 22, 48, and 99 nm (reproduced with permission from [66])

theory for noble metal particles such as Au and Ag. Qualitative estimates can however be easily made based on the observed plasmon resonance band.

Some of the features of plasmon bands and their dependence on the dielectric constant of the medium and the number of free electrons present in the nanocrystals find interesting uses. It is possible to probe events occurring in the vicinity of nanoparticles using the plasmon resonance band. There have been many experiments to demonstrate the high sensitivity of the plasmon band to small changes in the surrounding dielectric. Events that alter the surrounding dielectric such as binding of molecules and the solvent refractive index have been studied using plasmon resonance spectroscopy [66, 76, 77]. Chemisorption of molecules on the surface of the nanocrystals alters the number of free electrons as well as the dielectric constant of the medium, thereby leading to striking changes in the plasmon band [68].

In the case of semiconductor and other particles, the number of free electrons is much smaller and the plasmon absorption band is shifted to the infrared region. The absorption of visible radiation by semiconductor nanocrystals is due to excitonic transitions. Much of our understanding of the absorption processes in semiconductor nanocrystals in the visible region stems from the work of Efros and Brus and coworker [78–80]. They propose a theory based on effective mass approximation to explain the size-dependent changes in the absorption spectra of semiconducting nanocrystals. The absorption spectra can be understood by following the changes in the

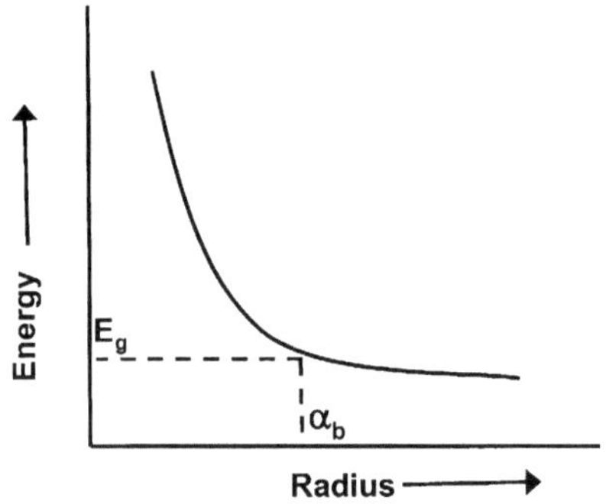

Fig. 1.17. A schematic illustration of the variation of absorption onset vs. the particle diameter. The absorption onset is directly related to the bandgap ($E_{\rm g}$) and is constant (bulk like) till the nanocrystal diameter is larger than the exciton diameter, $\alpha_{\rm b}$

size of the nanocrystals in comparison to the exciton diameter. It may be recalled that excitons are imaginary quasiparticles produced by pairing of an electron (from the conduction band) and a hole (from the valence band), due to Coulomb interaction. Excitons in semiconductors form hydrogen atom like states with diameters significantly larger than the unit cell length. The exciton diameter is, $\alpha_{\rm b}$, given by,

$$\alpha_{\rm b} = \frac{h^2\epsilon}{e^2}\left[\frac{1}{m_{\rm e}^*} + \frac{1}{m_{\rm h}^*}\right], \tag{1.22}$$

where $m_{\rm h}^*$ is the effective mass of a hole. When the diameter of a nanocrystals, D, is much larger than $\alpha_{\rm b}$, the absorption spectrum is similar to that of the bulk. However, when $\alpha_{\rm b}$ and D become closer, a sharp, size-dependent rise in the absorption onset accompanied by a blue-shift of the absorption maximum is seen. This is schematically illustrated in Fig. 1.17. Thus, CdS, a yellow solid, exhibits an excitonic absorption around 600 nm, which gradually shifts into the UV region as the nanocrystal is decreased till a value below 10 nm. In contrast to metal nanocrystals, the absorption band can be varied over a wide range by changing the dimensions of semiconductor nanocrystals. The excitonic nature of the absorption band permits a direct correlation of the bandgap of the semiconductor nanocrystal with the absorption edge. It is possible to model key effects such as the surface structure and coupling of the electronic states and reproduce the experimental observations [81,82]. However, our understanding of the optical properties is far from being complete.

In addition to properties associated with optical absorption, semiconductor nanocrystals exhibit interesting luminescent behavior. The luminescence is generally dependent on the size of the nanocrystal and the surface structure [83]. A photograph showing changes in the emission wavelength as a function of size of semiconductor quantum dots is shown in Fig. 1.18. The

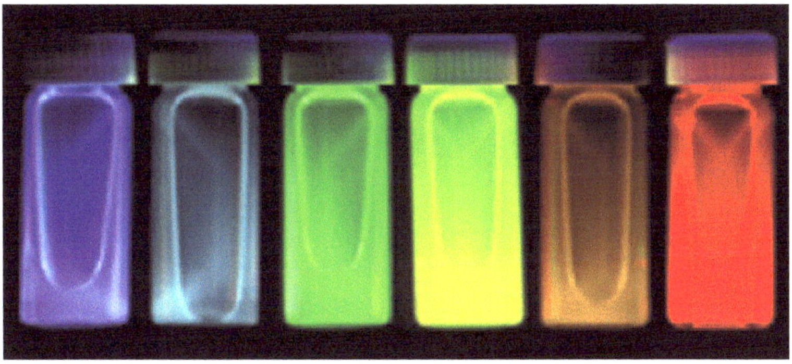

Fig. 1.18. Emission from ZnS coated CdSe nanocrystals of different sizes, dispersed in hexane (reproduced with permission from [83])

surface of a semiconductor nanocrystal can be quite different from that of the metal nanocrystals. Given the directional nature of the covalent bonds present in semiconductors, the surface structure cannot be described by schemes such as the ones indicated in (1.1). Growth defects tend to get concentrated at the surface, leading to surface states with energies in the midgap region, interfering with the emission process. It is, however, possible to obtain highly luminescent semiconductor nanocrystals by careful synthesis. A wide range of applications have been envisaged for luminescent nanocrystals.

1.2.5 Other Properties

An increase in the surface area per unit mass is a direct consequence of reduced dimensions of nanocrystals. The increase in specific surface area assumes significance in catalysis. Heterogenous catalysis benefits from the increase in the number of active sites accompanying the increase in surface area. Thus, nanoscale materials can be useful as powerful catalysts. Further, changes in the electronic structure brought about by quantum confinement effects could be used to tailor the reactivity of nanocrystals. Reactions with no parallels in bulk matter can be carried out through the aid of nanocrystals. A few such reactions have indeed been realized.

2

Synthesis of Nanocrystals

Modern materials science is characterized by a close interplay with physics, chemistry, and biology. This is specially true of nanomaterials, as vividly demonstrated by the methods of synthesis employed for these materials. On the one hand, are the top-down methods which rely on continuous breakup of bulk matter while on the other are the bottom-up methods that build up nanomaterials from their constituent atoms. The top-down and bottom-up approaches can also be considered as physical and chemical methods, respectively. A variety of hybrid methods have since come into being.

2.1 Physical Methods

Many of the physical methods involve the evaporation of a solid material to form a supersaturated vapor from which homogenous nucleation of nanoparticles occurs. In these methods, the size of the particles is controlled by temporarily inactivating the source of evaporation, or by slowing the rate by introducing gas molecules to collide with the particles. The growth generally occurs rapidly, from milliseconds to seconds, requiring a precise control over experimental parameters. Several specialized techniques have been developed in the last few decades and they can be classified on the basis of the energy source and whether they make use of solid or liquid (vapor) precursors [84]. Table 2.1 provides a summary of representative clusters and nanoparticles synthesized by physical methods.

2.1.1 Inert Gas Condensation

This method is most widely used and provides straightforward means to prepare nanosized clusters, specially of metals. A metal foil or ingot is heated in a ceramic crucible placed in a chamber filled with an inert gas, typically a few torr of argon. The metal vapor cools rapidly losing energy on collision with argon atoms, thereby producing nanoparticles. Thus, iron nanoparticles of

Table 2.1. Clusters and nanoparticles synthesized by physical methods (Representative examples from the literature)

system	method used	reference
Si	plasma, LP	[84, 102, 116–118, 120, 119]
Fe	IGC, IS	[85–87, 107]
Ni	plasma	[96]
Na	LA	[134]
Mg	IGC	[135]
Gd	IGC	[92]
Co	LP	[121]
Cu	SP	[127]
Ge	LP	[126]
Ag	LA	[112]
Au	IS	[133, 108]
U, Pu	LA	[109, 110]
Fe–Fe_2O_3(core–shell)	IGC	[88]
PbS–Ag	IGC	[93]
Al–Cu	IS	[137]
Al–(Mn, V, Cr)	IGC	[136]
CdS	IGC	[90]
CdSe	IS	[138]
CdTe	IS	[139]
ZnSe	IGC	[91]
MoS_2	LP	[122]
TiO_2	SP, LP	[128, 125]
VO_2	LA	[111]
SnO	IGC	[89]
SiO_2	SP	[131]
ZrO_2–SiO_2	plasma	[140]
Si_3N_4	plasma, LP	[100, 101, 123]
SiC	plasma, LP	[100, 101, 123]
WC	plasma	[100, 101]
TiC	plasma	[100, 101]
FeC	LP	[124]
AlN	plasma	[100, 101, 98, 97]
Ni–Mn–Fe ferrite	plasma	[103, 104]

IGC, inert gas condensation; plasma, arc, RF, or thermal; LP, laser pyrolysis; LA, laser ablation; SP spray pyrolysis; IS, ion sputtering.

5–40 nm diameter have been prepared by condensing in a helium or an argon atmosphere [85–87]. By controlled exposure of the particles to air, Baker et al. [88] converted the Fe particles to Fe–Fe_2O_3 core–shell structures. The particles were single domain but superparamagnetic, the blocking temperature varying with the thickness of the oxide shell. Lee et al. [89] have prepared SnO nanoparticles under different partial pressures of the convection gases, oxygen and helium. Semiconductor nanoparticles such as CdS [90] and ZnSe [91] have also been prepared using inert gas condensation. Gd nanoparticles condensed in an inert gas have been found to exhibit improved stability against oxidation [92]. A size-induced structural transition, from hcp to fcc, has been observed in these particles. Maisels et al. [93] advanced the technique to prepare composite nanoparticles of PbS and Ag brought together through separate channels. Nakaso et al. [94] studied the in-flight sintering of Au nanoparticles following condensation in the gaseous medium. Molecular dynamics simulations have been carried out to understand cluster nucleation under condensation conditions [95].

2.1.2 Arc Discharge

Another means of vaporizing metals is to strike an arc between metal electrodes in the presence of an inert gas. Weber et al. [96] used this method to prepare Ni nanoparticles and studied the in situ catalytic properties without interference from a substrate. Nanoparticles of metal oxides, carbides, and nitrides can be prepared by carrying out the discharge in a suitable gas medium or by loading the electrodes with suitable precursor (see Table 2.1). For example, Balasubramanian et al. [97] prepared cubic AlN particles (15–80 nm) by striking an arc between a W cathode and an Al anode in a gas mixture of nitrogen and argon. Iwata et al. [98] found it useful to blow a N_2–NH_3 mixture through the Al plasma to obtain AlN nanoparticles, on the walls of the suction pipe. Rexer et al. [99] developed a pulsed arc cluster deposition unit by replacing continuous flow by short gas pulses. Other related methods include thermal [100] and radio frequency (RF) [101, 102] plasma synthesis. The latter method has been used extensively to produce single crystalline Si nanoparticles [102]. Nanocrystalline ferrite powders (20–30 nm) of Ni–Mn–Fe have also been synthesized using RF plasma torch [103, 104]. Nanopowders of TiO_2 have been prepared by the plasma oxidation of Ti butoxide stabilized with diethanolamine [105].

2.1.3 Ion Sputtering

In this method, accelerated ions such as Ar^+ are directed toward the surface of a target to eject atoms and small clusters from its surface. The ions are carried to the substrate under a relatively high pressure (~1 mTorr) of an inert gas, causing aggregation of the species. Nanoparticles of metals and alloys as well as semiconductors have been prepared using this method (see Table 2.1). Urban et al. [106] have demonstrated the formation of nanoparticles of various metals using magnetron sputtering. They formed collimated beams of nanoparticles and deposited them as nanostructured films on Si substrates. Babonneau et al. [107] obtained spherical and elongated iron nanoparticles buried in carbon matrix. Birtcher et al. have made an interesting observation of Au nanoparticle formation from single Xe ion impacts [108].

2.1.4 Laser Ablation

Laser ablation is not only suited to metals but also to refractory materials, although the throughput may not be appreciable. A pulsed excimer laser or a Nd:YAG laser is generally used. A group at Lawrence Livermore National Laboratory has used this method to synthesize nanoparticles of actinide metals [109, 110]. Their instrumentation was particularly suitable to carry out in situ studies on highly toxic and radioactive materials, often in trace amounts. Another advantage of the laser ablation method is that the target composition is faithfully reproduced in the nanoparticles formed. For instance, Suh et al. [111] obtained nonstoichiometric $VO_{1.7}$ nanoparticles starting with target of a similar composition. The particles were subsequently oxidized to VO_2. A variant of the method is the supersonic expansion method where the plume produced by the laser pulse is carried by an inert gas pulse through a narrow orifice to cause adiabatic expansion. This results in the formation of nanoclusters. Harfenist et al. [112] prepared mass-selected Ag nanoparticles from this method and collected them in the form of a sol outside the preparation chamber. Because of spatial and temporal confinement of the gaseous species with supersonic speeds, the method calls for extreme sophistication in instrumentation [113–115].

2.1.5 Pyrolysis and Other Methods

In laser pyrolysis, a precursor in the gaseous form is mixed with an inert gas and heated with CO_2 infrared laser (continuous or pulsed), whose energy is either absorbed by the precursor or by an inert photosensitizer such as SF_6. Swihart [84], Ledoux et al. [116,117], and Ehbrecht and Huisken [118] prepared Si nanoparticles by laser pyrolysis of silane. By using a fast-spinning molecular beam chopper, Si nanoparticles in the size range of 2.5–8 nm were deposited on quartz substrates to study quantum confinement effects [116]. Li et al. [119] improved the stability of the Si nanoparticles ($\sim$5 nm) by surface functionalization and obtained persistent bright visible photoluminescence. Hofmeister et al. [120] have studied lattice contraction in nanosized Si particles produced by laser pyrolysis. The method has been used to synthesize metal nanoparticles as well (see Table 2.1). Zhao et al. [121] obtained Co nanoparticles by laser pyrolysis of $Co_2(CO)_8$ vapor at a relatively low temperature of 44°C. Ethylene was used as a photosensitizer for CO_2 laser emission. Nanoparticles of MoS_2 [122], Si_3N_4, and SiC [123] have also been prepared by this method. Grimes et al. [124] synthesized single phase ferromagnetic γ-Fe_3N and FeC nanoparticles and characterized them using a host of techniques. Nano-TiO_2 powders containing mixtures of anatase and rutile phases have been prepared by laser pyrolysis of $TiCl_4$ vapor, using air or nitrous oxide as oxygen carrier [125]. Ge nanoparticles have been deposited on a Si substrate using an organogermanium precursor [126]. By increasing the number of laser shots, the nanoparticles could be converted to Si–Ge alloy structures.

2.1.6 Spray Pyrolysis

In spray pyrolysis, small droplets of a solution containing a desired precursor are injected into the hot zone of a furnace to obtain nanoparticles. The droplets are generated by using a nebulizer, generally by making use of a transducer. By controlling the nebulizer energy, the relative vapor pressures of the gases and the temperature of the furnace, the particle size is controlled. Some examples of the use of spray pyrolysis are in the synthesis of Cu [127] and TiO_2 [128] nanoparticles. Starting with different Cd precursors, CdS and CdSe nanoparticles have been prepared [129, 130]. A variant of this technique is flame spray pyrolysis, where the heat needed for the decomposition of the precursor is produced in situ by combustion. Mädler et al. [131] applied this technique to synthesize SiO_2 nanoparticles from hexamethyldisiloxane.

Another physical method of preparing nanoparticles is high-energy ball milling. This method has been used extensively to produce metastable phases of alloys and intermetallics [132] as well as a variety of oxides and other materials.

2.2 Chemical Methods

Chemical methods have emerged to be indispensable for synthesizing nanocrystals of various types of materials. These methods are generally carried out under mild conditions and are relatively straightforward. Nanodimensional materials in the form of embedded solids, liquids, and foams have also been prepared by chemical means and such materials have been in use for some time. In the presentation that follows, we focus on the means of producing isolated nanocrystals dispersible in solvents (sols). There are several reviews in the literature focusing on the synthesis of nanocrystals [141–143].

Any chemical reaction resulting in a sol consists of three steps - seeding, particle growth, and growth termination by capping. An important process that occurs during the growth of a colloid is Ostwald ripening. Ostwald ripening is a growth mechanism whereby smaller particles dissolve releasing monomers or ions for consumption by larger particles, the driving force being the lower solubility of larger particles. Ostwald ripening limits the ultimate size distribution obtainable to about 15% of the particle diameter when the growth occurs under equilibrium conditions. However, by employing high concentrations of the monomers and capping agents, growth can be forced to occur in a transient regime. The seeding, nucleation, and termination steps are often not separable and one, therefore, starts with a mixture of the nanocrystal constituents, capping agents, and the solvent. The relative rates of the steps can be altered by changing parameters such as concentrations and temperature. This is the popular trick employed to obtain nanocrystals of different dimensions from the same reaction mixture.

One of the important factors that determine the quality of a synthetic procedure is the monodispersity of the nanocrystals obtained. It is desirable to have nanoparticles of nearly the same size, in order to be able to relate the size and the property under study. Hence, narrower the size distribution, more attractive is the synthetic procedure. The best synthetic schemes today produce nanocrystals with diameter distribution of around 5%. The other important issues are the choice of the capping agent and control over the shape.

Sols produced by chemical means can either be in aqueous media (hydrosols) or in organic solvents (organosols). Organosols are sterically stabilized, while hydrosols can either be sterically or electrostatically stabilized. Steric stabilization of hydrosols can be brought about by the use of polymers as stabilizing agents. Natural polymers such as starch and cellulose, synthetic polymers, such as polyvinyl pyrrolidone (PVP), polyvinyl alcohol (PVA), and polymethyl vinylether are used as stabilizing agents. Unlike text-book colloids such as India ink and dust in river beds, sterically stabilized sols are redispersible. The nanoparticulate matter in the sols can be precipitated by various means, filtered and dissolved again in a solvent. Redispersibility of the particles are an important characteristic of great utility. Furthermore, metal nanocrystals in a sterically stabilized sol can be dispersed in high concentrations. As a general rule, organic solvents provide better control over the size of the nanocrystals.

2.2.1 Metal Nanocrystals by Reduction

A variety of reducing agents are used to reduce soluble metal salts to obtain the corresponding metals. By terminating the growth with appropriate surfactants or ions, metal nanoparticles are produced. We shall discuss the use of a few representative reducing agents in this section. Some of the older methods of preparing nanoparticles were reviewed by Turkevich in 1951 [144].

Borohydride Reduction

Borohydride reduction, known since the 1950s, was one of the subjects of investigation under the Manhattan project [145, 146]. The basic reaction involves the hydrolysis of the borohydride accompanied by the evolution of hydrogen.

$$BH_4^- + 2H_2O \longrightarrow BO_2^- + 4H_2 \tag{2.1}$$

Nanocrystals of a variety of metals have been made by borohydride reduction [144]. Thus, Pt nanocrystals with mean diameter 2.8 nm were prepared by the reduction of chloroplatinic acid with sodium borohydride [147]. Homiyama and coworkers [148, 149] made Cu sols by the borohydride reduction of Cu salts. Green and O'Brien [150] prepared Cr and Ni nanoparticles by carrying out the reduction with Li or Na borohydride at high temperatures in coordinating solvents (see Fig. 2.1).

Schiffrin and coworkers [151] developed a two-phase method to reduce noble metals. This method, popularly known as the Brust method, has been widely used to prepare organosols. In this method, aqueous metal ions are transferred to a toluene layer by the use of tetraoctylammonium bromide, a phase transfer catalyst which is also capable of acting as a stabilizing agent. The Au complex transferred to toluene is reacted with alkanethiols to form polymeric thiolates. Aqueous borohydride is added to this mixture to bring about the reduction that is modulated by the interface of toluene and water (see Fig. 2.2). The thiol molecules also serve as capping agents. The capping action of the thiols is related to the formation of a crystalline monolayer on the

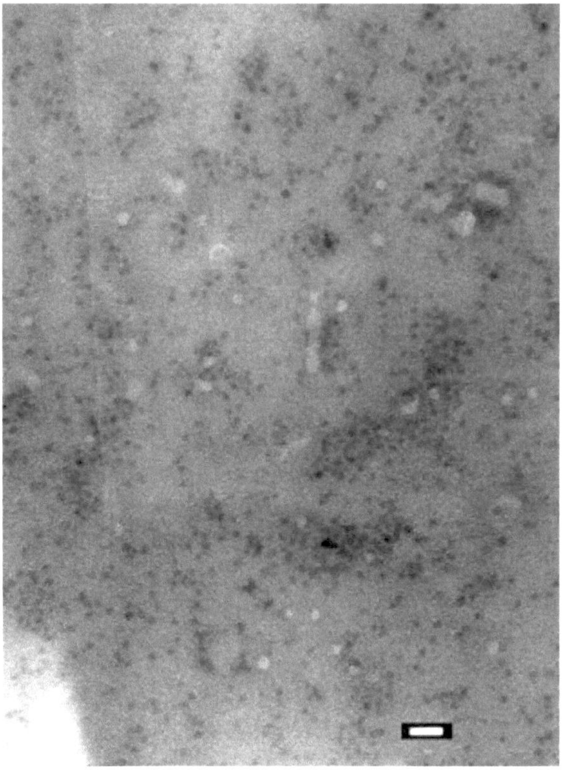

Fig. 2.1. TEM image of Cr nanocrystals synthesized by borohydride reduction. The scale bar corresponds to 20 nm (reproduced with permission from [150])

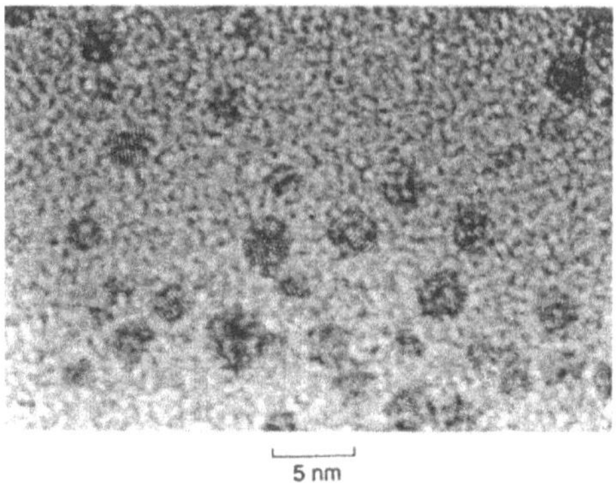

Fig. 2.2. TEM image of Au nanocrystals synthesized by the Brust method (reproduced with permission from [151])

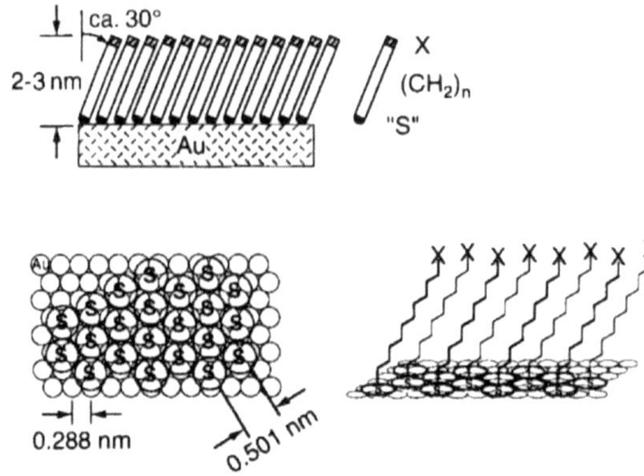

Fig. 2.3. Crystallographic details of the thiol adsorption on the Au(111) surface (reproduced with permission from [153])

metal particle surface (see Fig. 2.3). Since the formation of alkane thiol monolayers on Au was discovered by Nuzzo and Allara [152,153], the interaction of thiols with Au has been exploited widely in nanoscience. The powerful capping action of alkanethiols at metal particle surfaces prompted Murray [154] to name these particles monolayer protected clusters (MPC). The length of the alkane chain and the concentration of borohydride affect the size of the nanocrystals obtained by the Brust method. This method has been successfully extended to prepare MPCs of Ag and Pd as well [155–157].

Problems associated with borohydride reduction include its irreproducibility, specially in the aqueous medium and the incorporation of boron in the product [158,159]. Bönneman and coworkers [160,161] suggested the use of triethylborohydrides to avoid the incorporation of boron. In order to demonstrate the versatility of the borohydride reduction method, Bönneman and coworkers reduced a range of metal halides of early transition metals such as Ti, Zr, V,

Nb, and Mb using triethylborohydrides in tetrahydrofuran. Tetrahydrofuran also acts as the stabilizing agent in the reaction. By coupling tetraalkylammonium ions with triethylborohydride, a complex consisting of both the reducing agent and the stabilizing agent is obtained. Such a complex, when employed for reduction, yields small nanoparticles, due to the liberation of the stabilizing agent in the vicinity of the growing particle seed. By employing either of the two procedures, Bönneman and coworkers have prepared nanoparticles of metals belonging to groups 6–11. Sun and coworkers [162] have used lithium triethylborohydride ($LiBEt_3H$, also called superhydride) to reduce Co chloride in a solvent mixture consisting of oleic acid and an alkylphosphine. The alkylphosphine serves as a capping agent as well. The size of the nanocrystals could be tuned by varying the chain length of the alkylphosphine. Cobalt nanoparticles of 3 nm diameter have also been prepared using di-isobutyl aluminum hydride [163].

Citrate Reduction

The citrate route to colloidal Au, first described by Hauser and Lynn [164], is a popular method to prepare Au hydrosols. Synthesis by the citrate method involves the addition of chloroauric acid to a boiling solution of sodium citrate [144]. A wine red color indicates the onset of reduction. The average diameter of the nanoparticles can be varied over a range of 10–100 nm by varying the concentration ratio between chloroauric acid and sodium citrate (see Fig. 2.4). Turkevich [165] proposed that the reaction involved the formation of acetone dicarboxylate. Subsequent reduction occurs by the dicarboxylate species. The citrate method has been extended to synthesize of Pt and Ag nanoparticles [166–168]. Turkevich and coworkers [169] prepared Au–Pt

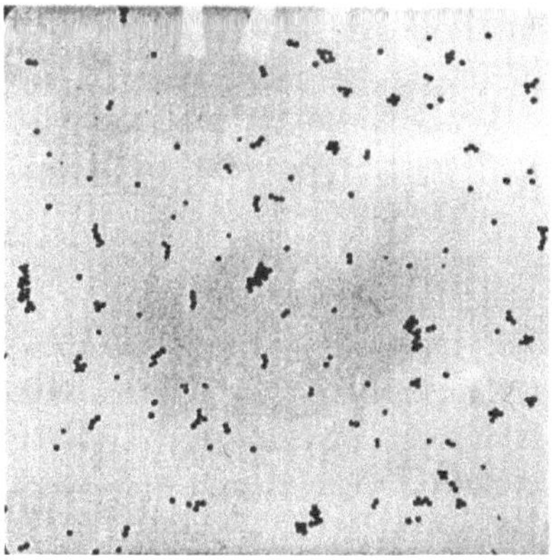

Fig. 2.4. TEM image of Au nanocrystals synthesized by citrate reduction. The magnification is 50,000 (reproduced with permission from [144])

bimetallic nanoparticles by reducing a mixture of chloroauric acid and chloroplatinic acid by citrate. By a similar procedure, Pd–Pt nanoparticles could also be prepared.

Alcohol Reduction

The ability of small metal particles in catalyzing the oxidation of alcohols to aldehydes or acids has long been known in organic chemistry. The fine metal particles that catalyse these reactions get reduced during the course of the reaction and are deposited as spongy precipitates. This reaction is further catalyzed by a base and requires the presence of α-hydrogen in the alcohol. By making use of polymeric capping agents such as PVP, the growth of metal particles can be arrested [170–172]. The mechanism of the reaction involves the formation of an alkoxide, with an oxonium ion as the intermediate, followed by the formation of a carbonyl compound by hydride elimination. The base aids the deprotonation of the oxonium intermediate. A counter ion, such as the acetate, serves as an effective base. Otherwise, water is added, specially in the case of lower alcohols. Palladium acetate refluxed with ethanol in the presence of PVP yields Pd nanocrystals of 6.0 nm diameter [173–175]. Organosols of Pd are formed by the reduction of Pd salts in 1-decanol. In this case, the alcohol also acts as the stabilizing agent [176]. In an elegant demonstration of size control, Teranishi and Miyake [177] have prepared monodisperse PVP-covered Pd nanocrystals of various sizes in the 2–5 nm range, by employing ethanol as the reducing agent and varying its power by diluting with water. Larger Pd nanocrystals have been prepared by employing the small nanocrystals as seeds in the reaction mixture. Nanoparticles of Ag, Au, Pd, and Cu have been prepared by ethyl alcohol reduction of metal salts under refluxing conditions in the presence of PVP [177, 178]. Other metal sols prepared by alcohol reduction include Pt, Au, and Rh [179]. Toshima and coworkers [180, 181] have prepared PVP-capped Pd–Pt alloy nanoparticles by reducing a mixture of palladium chloride and chloroplatinic acid in a alcohol–water mixture. Ag–Pd and Cu–Pd alloy nanoparticles of a wide range of composition have been prepared by the alcohol reduction method [182].

The polyol method of Figlarz and coworkers [183, 184] involves the reduction of metal salts with high boiling polyols such as ethylene glycol. The polyol serves both as the reducing agent and the stabilizing agent. Particles of Co, Ni, Cu, Au, Ag, and their alloys in the size range of 100 nm to a few microns have been obtained by this method (see Fig. 2.5) [185–188]. The polyol method provides a convenient chemical route to nanoparticles in this size range, which are otherwise difficult to obtain. By employing PVP as the capping agent and a series of polyols as reducing agents, Liu and coworkers [189] have made small Ru particles in the size range of 1.4–7.4 nm. By simultaneously reducing a mixture of Cu and Pd hydroxides (obtained by hydrolysis of Cu sulphate and Pd acetate in glycol), Toshima et al. [190–192] obtained PVP-capped Cu–Pd bimetallic nanoparticles with an average diameter of 2 nm. Ni–Pd

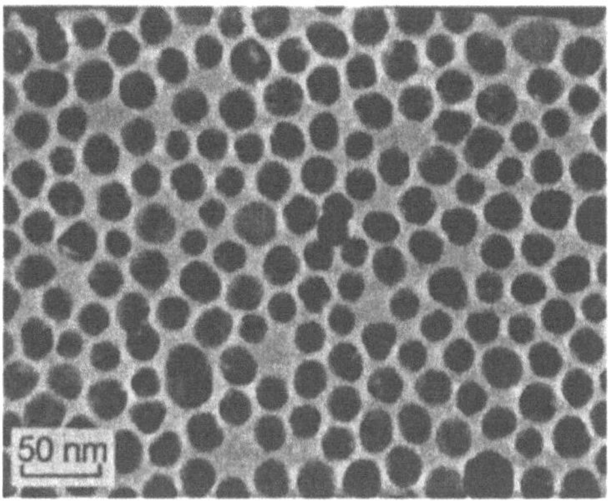

Fig. 2.5. TEM image of large Ag nanocrystals synthesized by the polyol method (reproduced with permission from [185])

Table 2.2. Metal particles synthesized by alkylaluminate reduction

metal salt	reducing agent	T (°C)	time (h)	d (nm)
$PtCl_2$	$Al(Me)_3$	40	16	2.0
$Pt(acac)_2$	$Al(Me)_3$	20	24	2.5
Ag neodecanoate	$Al(Oct)_3$	20	36	8–12
$Pd(acac)_2/Pt(acac)_2$	$Al(Et)_3$	20	2	3.2
$Pt(acac)_2/Ru(acac)_2$	$Al(Me)_3$	60	21	1.3

and Ni–Pt nanoparticles could also be prepared by hydrolysis followed by polyol reduction [193]. The polyol process has also been employed to obtain PtBi nanoparticles [194]. Bradley and coworkers [195] have prepared Cu–Pd bimetallic nanoparticles by decomposing a mixture of Cu and Pd acetates in a high boiling alcohol (ethoxyethanol).

Other Reducing Agents

Alkylaluminates (AlR_3, R=C_1–C_8) are used as reducing agents for reducing metals such as Ni, Fe, Ag, and Pd [196]. The reactions are typically carried out in toluene at temperatures in the 20–60°C range. The alkylaluminate also acts as the stabilizing agent. Bimetallic nanoparticles could also be prepared by this method. Table 2.2 lists the conditions and reagents used in alkylaluminate reductions.

Alkalides and electrides are effective reducing agents comparable to solvated electrons. Alkalides and electrides are crystalline salts consisting of crown ethers complexed with alkali ions or salts with alkali metals as anions, and consist of trapped electrons. They are of the type $K^+(15-\text{crown}-6)_2Na$ [197–200]. The reduction is carried out in solvents such as THF under

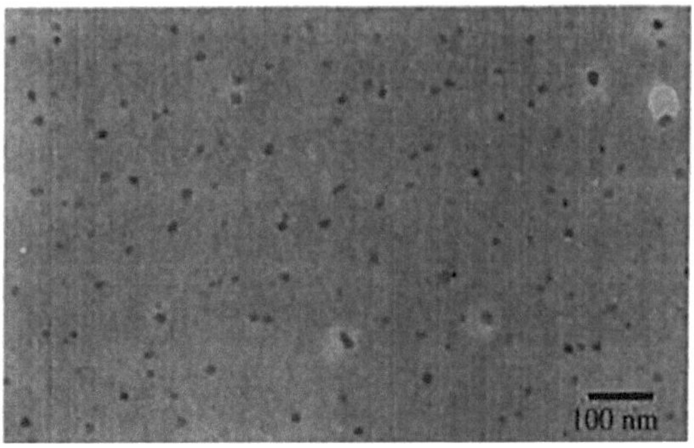

Fig. 2.6. TEM image of oxide nanocrystals obtained upon oxidation of Gd nanocrystals produced by alkalide reduction (reproduced with permission from [201])

scrupulously dry conditions. Au nanocrystals are produced by the alkalide reduction of $AuCl_3$. The familiar wine red hydrosol is produced on dissolving the product in water [197]. Wagner and coworkers [201, 202] have made Gd and Dy nanocrystals by using alkalides in tetrahydrofuran (see Fig. 2.6).

Closely related to the alkalide reduction method is the reduction by alkali metals such as K, Li, and Na. In dry THF, diglyme or other ethers, metal halides are reduced to obtain nanoparticles of metals. $NiCl_2$ has been reduced to Ni in diglyme while $AlCl_3$ has been reduced to Al in xylene by Rieke and coworkers [203, 204]. An important extension to this method is the use of the alkali naphthalides [205].

Reduction carried out by hydrogen and gases such as CO/H_2O (water gas) and diborane provides certain advantages over other reducing agents. Gaseous reductions are clean and seldom leave by-products in the reducing mixture. At the end of the reaction, the remaining traces of the reducing agent are removed by warming or airing the mixture. Pd nanocrystals are prepared by reducing Pd acetate with H_2 in acetic acid in the presence of 1,10-phenanthroline [206]. Schmid and coworkers [207] have prepared Au nanocrystals by reducing gold triphenylphosphonium chloride ($Au(PPh_3)Cl$) with diborane in dichloromethane. CO/H_2O has been used to reduce PtO_2 to Pt [208].

Exceedingly fine Au nanoparticles with diameters between 1 and 2 nm have been made by reduction of $HAuCl_4$ with tetrakishydroxymethylphosphonim chloride (THPC)(see Fig. 2.7) [209]. THPC also acts as a capping agent. Similarly, Au nanoparticles, free from surfactants, have been synthesized by the reduction of the aurate ion by sodium napthalenide in diglyme [210]. Liz-Marźan and coworkers [211, 212] have prepared nanoscale Ag nanoparticles by using dimetylformamide (DMF) as both a stabilizing agent and a capping agent. Cu and Pt nanoparicles have been made by using hydrazine hydrate as reducing agent [213, 214]. By using tetrabutylammonium borohydride or its mixture with hydrazine, Peng and coworkers [215] have obtained monodisperse nanocrystals of Au, Cu, Ag, and Pt. In this method, metal

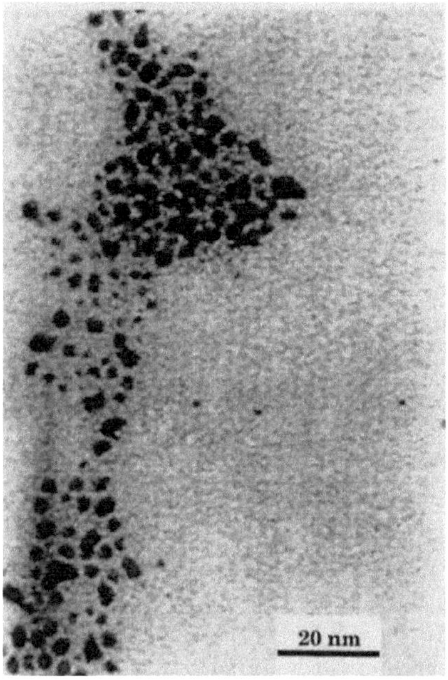

Fig. 2.7. TEM image of small Au nanocrystals obtained by reduction with tetrakishydroxymethylphosphonim chloride as reducing agent (reproduced with permission from [209])

ions from $AuCl_3$, $Ag(CH_3COO)$, $Cu(CH_3COO)_2$, and $PtCl_4$ were dispersed in toluene with the aid of long-chain quaternary ammonium salts and reduced with tetrabutylammonium borohydride, a toluene soluble borohydride. This reaction can be scaled up to produce gram quantities of nanocrystals.

Reductive Synthesis of Other Nanocrystals

Naphtalides, alkalides, and alkali metals are sufficiently powerful to reduce Ge and Si salts to the elements. Si nanocrystals have been prepared in solution by the reduction of the halides with Na, Li naphthalide, and hydride reagents [216–219]. Similarly, Ge nanocrystals have been made by the reduction of $GeCl_4$ with Li naphthalide in THF [217]. TEOS ($Si(OEt)_4$) is readily reduced by sodium to yield Si nanocrystals. Si and Ge nanocrystals are frequently covered by an oxide layer. Y_2O_3 nanocrystals have been made by the alkalide reduction of YCl_3 followed by oxidation by exposure to ambient conditions [220]. Yittria nanocrystals could be doped with Eu to render them phosphorescent [221]. ZnO nanoparticles have been prepared from zinc acetate in 2-propanol by the reaction with water [222]. Pure anatase nanocrystals are obtained by the hydrolysis of $TiCl_4$ with ethanol at 0°C followed by calcination at 87°C for 3 days [223]. The growth kinetics and the surface hydration chemistry in this reaction have been investigated.

2.2.2 Solvothermal Synthesis

The solvothermal method provides a means of using solvents at temperatures well above their boiling points, by carrying out the reaction in a sealed vessel. The pressure generated in the vessel due to the solvent vapors elevates the boiling point of the solvent. Typically, solvothermal methods make use of solvents such as ethanol, toluene, and water, and are widely used to synthesize zeolites, inorganic open-framework structures, and other solid materials. Due to the high-pressures employed, one often obtains high-pressure phases of the materials. In the past few years, solvothermal synthesis has emerged to become the chosen method to synthesize nanocrystals of inorganic materials. Numerous solvothermal schemes have been employed to produce nanocrystalline powders as well as nanocrystals dispersible in a liquid [224].

Qian and coworkers [225] have reported several solvothermal routes to chalcogenide nanocrystals. CdS nanocrystals of 6 nm diameter have been made using cadmium sulphate/nitrate as the Cd source, thiourea as the S source and ethyleneglycol as the solvent. The reaction was carried out for 12 h at 180°C. Chen and Fan [226] have prepared transition metal dichalcogenides (MS_2; M = Fe, Co, Ni, Mo; S = S or Se) with diameters in the range 4–200 nm by a hydrothermal route (water as solvent). Fe, Co, and Ni chalcogenides were obtained by treating the corresponding halide with $Na_2S_2O_3$ (sodium thiosulphate) or Na_2SeSO_3 (sodium selenosulphate) for 12 h at 140–150°C. Mo chalcogenides were prepared starting from Na_2MoO_4, sodium thio or seleno sulphate and hydrazine.

By employing a metal salt, elemental Se or S and a reducing agent (to reduce Se or S), it is possible to produce metal chalcogenide nanocrystals (see Table 2.3). Control over size is rendered possible by the slow release of sulphide or selenide ions. Nanocrystal dispersions are often obtained even without a capping agent. In some cases, S or Se can be caused to disproptionate, making the reducing agent redundant. Thus, CdSe nanocrystals have been prepared solvothermally by reacting Cd stearate with elemental Se in toluene in the presence of tetralin (see Fig. 2.8) [227]. The key step in the reaction scheme is the aromatization of tetralin to naphthalene in the presence of Se, producing H_2Se. CdS nanocrystals are similarly prepared by the reaction of a Cd salt with S in the presence of tetralin. Wei et al. report a green synthetic route for CdSe nanoparticles in aqueous solution using Se powder as the selenium source [228]. CdS nanocrystals of ~6 nm diameter have also been prepared in an aqueous solution containing pepsin at room temperature [229].

Nanocrystals of chalcogenides can be prepared under solvothermal conditions starting with the elements. Thus, ZnSe nanocrystals of 12–16 nm are prepared by reacting the elements in pyridine. By reacting Zn and Se in ethylenediamine, a complex ZnSe(en) is obtained [240]. This compound gets thermally decomposed to yield ZnSe nanoparticles. Interestingly, ZnSe particles posses the wurzite structure, instead of the normal cubic structure. Peng et al. [241] have obtained ZnSe and CdSe nanoparticles in the

Table 2.3. Chalcogenide nanocrystals obtained by solvothermal reactions involving reduction

salt, chalcogenide source	reducing agent, solvent	*d* (nm)	reference
Cd stearate, Se	toluene	3	[227]
CuI, Se	ethylene diamine	18	[230]
$Co(ac)_2$, thiourea	hydrazine, water	6	[231]
$SnCl_2$, Se	Na, ethylene diamine	45	[232]
$BiCl_3$, Se	NaI, ethylene diamine	25	[233]
$InCl_3$, Na_2S_2 3	ethanol	15	[234]
MCl_2 (M=Fe, Co, Ni), Se	hydrazine, H_2O/DMF		[235]
MCl_2 (M=Fe, Co, Ni, Mo), $Na_2S_2O_3$/ Na_2SeSO_3	water		[226]
$CuCl_2$, $InCl_3$, Se	diethylamine	15	[236]
$CuCl_2$, $InCl_3$, S/Se	ethylenediamine	15	[237]
AgCl, In/Ga, S	ethylenediamine	6,10	[238]
CuCl, In/Ga, S	water	10,35	[239]

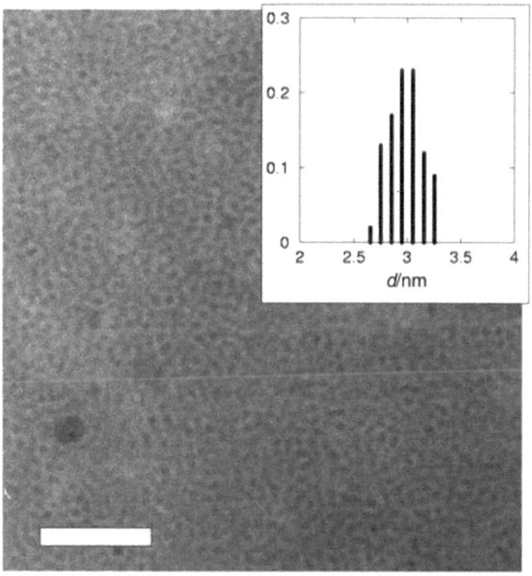

Fig. 2.8. TEM image of CdSe nanocrystals synthesized by solvothermal method. The scale bar corresponds to 50 nm. The inset shows a histogram of particle size distribution (reproduced with permission from [227])

70–100 nm size range by combining the elements under hydrothermal conditions. ZnS nanoparticles in the 3–18 nm size range have been obtained by Qian and coworkers [242] by treating Zn and S in pyridine. A surfactant-assisted solvothermal procedure has been employed to prepare PbS nanocrystals at

85°C. Other nanostructures are also obtained by this preparation [243]. A solvothermal reaction in the presence of octadecylamine yields monodisperse PbSe nanocrystals of controllable size [244]. A low-temperature one-pot synthesis of HgTe nanocrystals has been described without the use of toxic precursors. The particles show infrared photoluminescence [245].

Large GaN nanocrystals (32 nm) have been produced by Xie et al. by treating $GaCl_3$ with Li_3N in benzene [246]. Sardar and Rao [247] have prepared GaN nanoparticles of various sizes under solvothermal conditions, employing gallium cupferronate ($Ga(C_6H_5N_2O_2)_3$) or chloride ($GaCl_3$) as the gallium source and hexamethyldisilyzane as the nitriding agent and toluene as solvent (see Fig. 2.9). By employing surfactants such as cetyltrimethylammonium bromide (CTAB), the size of the nanocrystals could be controlled. In the case of cupferronate, the formation of the nanocrystals is likely to involve nitridation

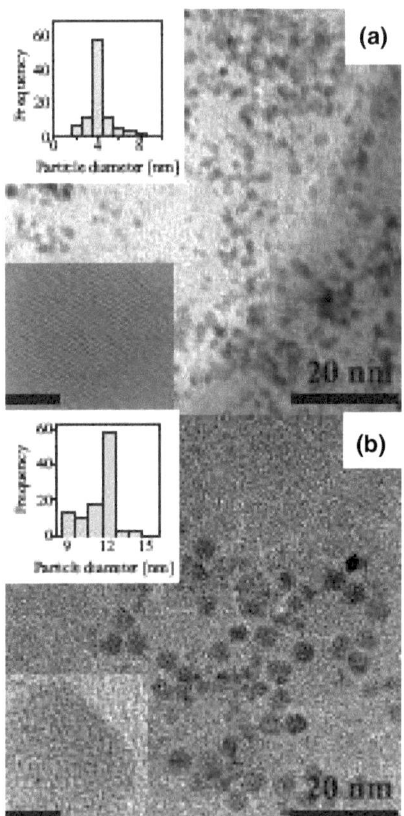

Fig. 2.9. TEM images and histograms showing the distribution in diameters of GaN nanocrystals produced by solvothermal decomposition of gallium cupferronate. HRTEM images are shown as insets. The scale bars correspond to 2.5 and 3.0 nm, respectively (reproduced with permission from [247])

of the nascent gallium oxide nanoparticles formed by the decomposition of the cupferronate.

$$GaO_{1.5} + (CH_3)_3SiNHSi(CH_3)_3 \longrightarrow GaN + (CH_3)_3SiOSi(CH_3)_3 + 1/2H_2O \quad (2.2)$$

With $GaCl_3$ as precursor, the reaction is,

$$GaCl_3 + (CH_3)_3SiNHSi(CH_3)_3 \longrightarrow GaN + Si(CH_3)_3Cl + HCl \quad (2.3)$$

This method has been applied for the synthesis of AlN, GaN, and InN nanocrystals [248]. The procedure yields nanocrystals with an average diameter of 10 nm for AlN, 15 nm for InN and as low as 4 nm for GaN.

InP nanoparticles of different sizes (12–40 nm) have been prepared by the solvothermal reaction involving $InCl_3$ and Na_3P [249]. The size of the particles could be varied by changing the solvent from coordinating dimethoxyethane to noncoordinating benzene. Semiconducting phosphides such as InP can be prepared by using the organometallic reagent $(CH_3)_3SiPhSi(CH_3)_3$ in a manner analogs to reaction (2.3) earlier.

Biswas and Rao [250] have prepared metallic ReO_3 nanocrystals of different sizes by the solvothermal decomposition of the rhenium(VII)oxide-dioxane complex, $(Re_2O_7\text{–}(C_4H_8O_2)_x)$, in toluene (see Fig. 2.10). The diameter of the nanocrystals could be varied in the range of 8.5–32.5 nm by varying the decomposition conditions. The reaction of metal acetylacetonates under solvothermal conditions produces nanocrystals of metal oxides

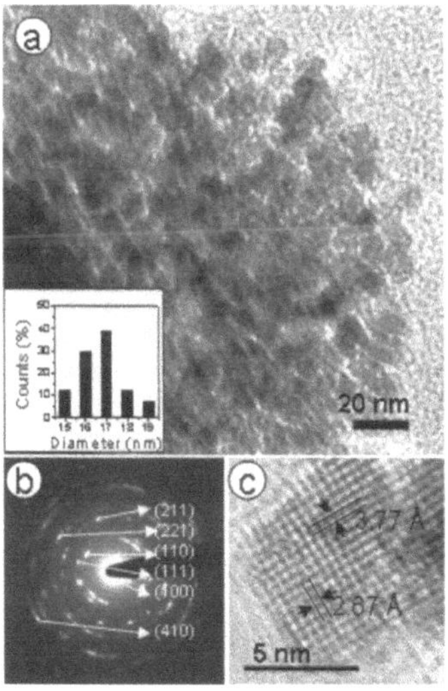

Fig. 2.10. (**a**) TEM image of ReO_3 nanoparticles with diameter of 17.0 nm (**b**) high-resolution TEM image of a 8.5 nm nanocrystal showing the (100) and (111) planes of ReO_3 (**c**) electron diffraction pattern obtained from ReO_3 nanoparticles (reproduced with permission from [250])

such as Ga_2O_3, ZnO, and cubic In_2O_3 [251]. CoO nanoparticles with diameters in 4.5–18 nm range have been prepared by the decomposition of cobalt cupferronate in decalin at 270°C under solvothermal conditions. Magnetic measurements indicate the presence of ferromagnetic interaction in small CoO nanoparticles [44]. Cubic and hexagonal CoO nanocrystals have also been obtained starting from $Co(acac)_3$ [252]. Nanoparticles of MnO and NiO have been synthesized from cupferronate precursors under solvothermal conditions. These nanoparticles exhibit superparamagnetism accompanied by magnetic hysteresis below a blocking temperature [45]. Metal nanocrystals have also been obtained by solvothermal synthesis. For example, nickel acetate gives metal nanoparticles in alkyl amine media; nanoparticles of Ru, Rh, and Ir can be obtained similarly by treating the acetylacetonates in the amine solvents under solvothermal conditions (see Sect. 2.2.6).

2.2.3 Photochemical Synthesis

Photochemical synthesis of nanoparticles is carried out by the light-induced decomposition of a metal complex or the reduction of a metal salt by photogenerated reducing agents such as solvated electrons. The former is called photolysis and the latter radiolysis. The formation of photographic images on a AgBr film is a familiar photolysis reaction. Henglein, Belloni, and their coworkers have pioneered the use of photolysis and radiolysis for the preparation of nanoscale metals [253]. Metals such as Cd and Tl have been obtained by photolysis. PVP-covered Au nanocrystals are produced by the reduction of $HAuCl_4$ in formamide by UV-irradiation [254]. The reaction is free radical mediated, with the radicals being generated by photodegradation of formamide. This provides a route to ion-free reduction of $HAuCl_4$. Radiolysis of Ag salts in the presence of polyphosphates produces extremely small clusters that are stable in solution for several hours. Effective control can be exercised over the reduction process by controlling the radiation dosage. Marandi et al. [255] have shown that the size of CdS nanocrystals could be controlled photochemically in the reaction of $CdSO_4$ and $Na_2S_2O_3$. Radiolysis also provides a means for the simultaneous generation of a larger number of metal nuclei at the start of the reaction, thereby yielding a fine dispersion of nanocrystals. Studies of the reduction pathways by radiolysis have been carried out [256].

2.2.4 Electrochemical Synthesis

Reetz and coworkers [257–265] have pioneered the electrochemical synthesis of metal nanocrystals. Their method represents a refinement of the classical electrorefining process and consists of six elementary steps they are oxidative dissolution of anode, migration of metal ions to the cathodes, reduction of ions to zero-valent state, formation of particles by nucleation and growth, arrest of growth by capping agents, and precipitation of particles. The steps

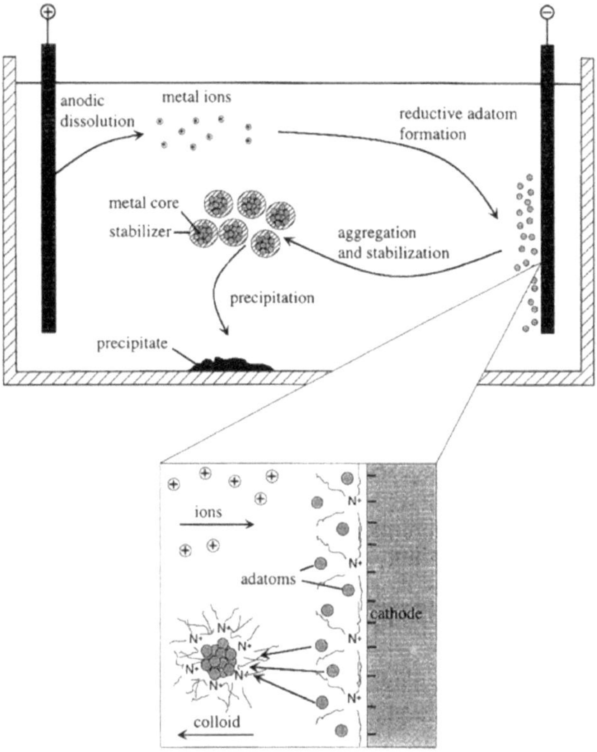

Fig. 2.11. Schematic illustration of the steps involved in the electrochemical reduction of metal nanocrystals by the Reetz method (reproduced with permission from [265])

are schematically illustrated in Fig. 2.11. The capping agents are typically quaternary ammonium salts containing long-chain alkanes such as tetraoctylammonium bromide. The size of the nanocrystals could be tuned by altering the current density, the distance between the electrodes, the reaction time, the temperature, and the polarity of the solvent. Thus, using tetraoctylammonium bromide as stabilizer, Pd nanocrystals in the size range of 1–5 nm have been obtained [257]. Low current densities yield larger particles (~4.8 nm) while large current densities yield smaller particles (~1.4 nm). Larger Pd nanoparticles stabilized by the solvent (propylene carbonate) have also been obtained [259]. This method has been used to synthesize Ni, Co, Fe, Ti, Ag, and Au nanoparticles. Bimetallic colloids such as Pd–Ni, Fe–Co, and Fe–Ni have been prepared using two anodes consisting of either metals [260]. Mono and bimetallic particles consisting of Pt, Rh, Ru, and Mo could be prepared by reduction of their salts dissolved in the electrolyte (see Table 2.4) [261]. Bimetallic particles could be prepared by using two ions, one of which was from the anode and the other from the metal salt dissolved in the electrolyte

Table 2.4. Metal particles synthesized by the electrochemical reduction of salts

metal salt	*d* (nm)
$PtCl_2$	2.5
$PtCl_2$	5.0
$RhCl_3H_2O$	2.5
$RuCl_3H_2O$	3.5
$OsCl_3$	2.0
$Pd(OAc)_2$	2.5
$Mo_2(OAc)_4$	5.0
$PtCl_2$+ $RuCl_3H_2O$	2.5

Table 2.5. Bimetallic particles synthesized by the combination of anodic oxidation and salt reduction

anode	metal salt	*d* (nm)
Sn	$PtCl_2$	3.0
Cu	$Pd(OAc)_2$	2.5
Pd	$PtCl_2$	3.5

(see Table 2.5) [261]. Pascal et al. [266] synthesized maghemite nanocrystals in the size range of 3–8 nm by the use of an Fe electrode in an aqueous solution containing DMF and cationic surfactants.

2.2.5 Nanocrystals of Semiconductors and Other Materials by Arrested Precipitation

Nanocrystals can be obtained from solutions that precipitate the bulk matter under conditions unfavorable for the growth of particulates in the precipitate. For example, the precipitation of metal salts by chalcogens can be arrested by employing a high pH. The groups of Brus, Henglein, and Weller have prepared CdS nanocrystals by adopting this strategy [267–269]. Typically, $CdSO_4$ is reacted with $(NH_4)_2S$ in water at high pH to obtain CdS particles of diameter around 5 nm. Other sulfur sources such as H_2S and Na_2S are also used to obtain CdS. Capping agents (e.g., sodium polyphosphate) stabilize such dispersions. In addition to water, methanol, acetonitrile and such solvents can be used to obtain CdS and ZnS nanocrystals by arrested precipitation [270]. Better control over particle size is attained by the use of stronger capping agents. Preparation of water soluble thiolate capped PbS nanocrystals has been investigated by Cornacchio et al. [271]. They also give a top-down protocol for the synthesis of PbS nanocrystals starting with bulk PbS dissolved in aqueous solutions of thioglycerol. Weller and coworkers [272–277] have pioneered the use of water soluble thiols, such as 1-thioglycerol, 2-mercaptoethanol, 1-mercapto-2-propanol, 1,2-dimercapto-3-propanol, thioglycolic acid, thiolactic acid, and cysteamine as capping agents to prepare CdS, CdSe, CdTe, HgSe, HgTe, and CdHgTe nanocrystals. Typically, a solution containing a metal salt (e.g., perchlorate) and the capping agent is treated with NaOH to raise the pH, degassed by bubbling inert gas (to prevent the oxidation of chalcogen source), followed by the introduction of the chalcogen in the form of Na_2S, NaHSe, etc. under inert conditions (see Fig. 2.12). Similarly, Zn_3P_2 and Cd_3P_2 nanocrystals have been prepared by the injection of phosphine gas into solutions of the corresponding perchlorates [278, 279].

Nanocrystals of metal oxides have been prepared by arresting the growth of precipitates formed under hydrolytic conditions. pH plays a key role in these reactions, a high pH being essential to precipitate basic oxides to scavenge the large number of liberated protons while amphoteric oxides normally require low pH. By controlling the pH, control is exercised over the size of particulates in the precipitate. Klabunde and coworkers [280] have prepared $MnFe_2O_4$ nanoparticles by the precipitation of the corresponding salts at a high pH. γ-Fe_2O_3 and $CoFe_2O_4$ nanocrystals are also obtained by this method [281]. By changing the base and temperature the size of the nanocrystals could be varied. ZnO nanocrystals have been prepared by the hydrolysis of zinc acetate in alcoholic solutions [282–284]. The use of long-chain alcohols and polyols in hydrolytic reactions provides certain advantages. These solvents dissolve both the metal salt and the organic capping agent and support hydrolysis reactions as well. Adopting a procedure similar to

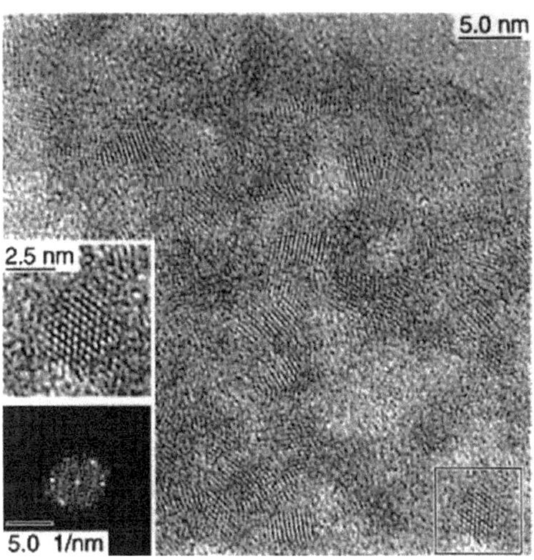

Fig. 2.12. TEM images of thioglycerol–capped CdSe nanocrystals prepared by arrested precipitation reaction. Insets show a HRTEM image and a Fourier transform of one of the HRTEM images (reproduced with permission from [272])

the polyol method, Ammar et al. [285] prepared slightly elongated $CoFe_2O_4$ nanoparticles. Carunto et al. [286] prepared a range of transition metal ferrites capped with long-chain carboxylic acids, using diethyleneglycol as the solvent.

2.2.6 Thermolysis Routes

Thermolysis routes are related to chemical vapor deposition (CVD)-based methods to prepare thin films. By carrying out thermolysis reactions in high boiling solvents in the presence of capping agents, nanocrystals of various materials are obtained. Thermal decomposition provides remarkable control over size and is well suited for scale up to gram quantities.

Metal and Metal Oxide Nanocrystals

Various metal nanoparticles have been prepared by decomposition of low-valent complexes involving olefinic ligands, such as cyclooctatetraene (COT), cycloocta-1,5-diene (COD), and carbonyls. It has been known since long that colloidal Co can be prepared by the decomposition of Co carbonyls [287]. Bawendi and coworkers [288] carried out a similar reaction with $Co_2(CO)_8$, in the presence of tri-*n*-octylphosphine oxide (TOPO) and obtained Co nanoparticles with an average diameter of 20 nm. By using capping agents such as carboxylic acids and alkyl amines the size of the nanoparticles can be tuned to be in the range of 3–20 nm [289–293]. Decomposition of carbonyls has been used to prepare nanocrystals of Fe [294, 295], FeCo [294], FeMo [296], FePt [297], CoPt [298], FePd [298], and SmCo [299] as well. Large Au nanoparticles with diameters of tens of nanometers have been prepared by Nakamoto et al. [300] by the thermolysis of Au(I) thiolate complexes – $[R(CH_3)_3N][Au(SC_{12}H_{25})_2]$, $[R(CH_3)_3N][Au(SC_6H_4\text{-p-}R')_2]$ $(R=C_{14}H_{29}, C_{12}H_{25})$; $(R'=C_8H_{17}, CH_3)$(see Fig. 2.13).

Chaudret and coworkers [301–303] have formulated a general method for the synthesis of various noble and magnetic metal nanoparticles. The method involves the reduction of low-valent olefinic complexes with H_2 and CO. Ag, Pt, Ru, Rh, Cu, Ir, Zn, PtRu, Co, and Ni nanoparticles have been prepared by this method. Yase and coworkers [304] have synthesized Ag nanoparticles by the thermal decomposition of Ag carboxylates. Quite remarkably, the reaction carried out in solid state yields capped Ag nanoparticles dispersible in toluene. This method is very interesting from the point of view of scale-up. Yase and coworkers have adopted this method to prepare conducting silver patterns by first screen printing with a paste consisting of Ag carboxylates. Kim et al. [305] have prepared Pd nanoparticles by decomposing Pd acetate in a medium made of TOPO/TOP and oleylamine. Nanocrystals of Cu have been obtained by thermolysis of a CVD precursor $Cu(OCH(CH_3)CH_2N(CH_3)_2)$ [306]. Thermolytic reactions of $Pt(NH_3)_2Cl_2$ in different high boiling solvents give rise to Pt nanocrystals of complex morphologies [307].

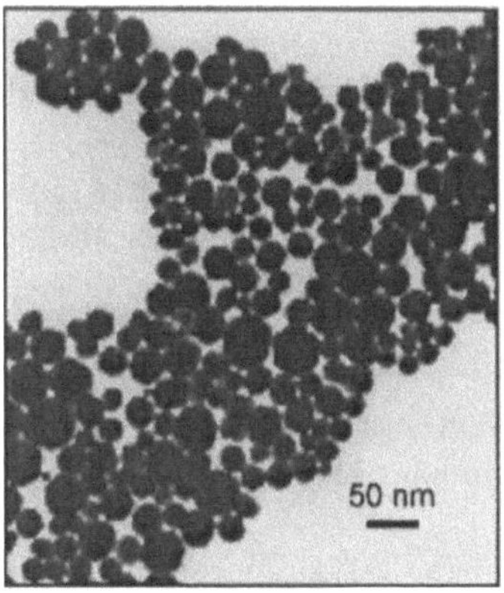

Fig. 2.13. TEM image of large Au nanocrystals prepared by decomposition of Au thiolate $[C_{14}H_{29}(CH_3)_3N][Au(SC_{12}H_{25})_2]$ (reproduced with permission from [300])

Iron–platinum alloy nanocrystals have attracted attention due to their magnetic properties [308]. Fe–Pt nanocrystals with 1:1 ratio of Fe:Pt exist in two forms. An fcc (face centered cubic) form in which the Fe and Pt atoms are randomly distributed (A1 phase) or an fct (face centered tetragonal) form in which Fe and Pt layers alternate along the $\langle 001 \rangle$ axis ($L1_0$ phase). The latter phase has the highest anisotropic constant among all known magnetic material. Synthesis of homogeneously alloyed Fe–Pt nanocrystals requires that Fe and Pt are nucleated at the same time. This process was first accomplished by reducing platinum acetylacetonate with long-chain diol and decomposing $Fe(CO)_5$ in the presence of oleic acid and a long-chain amine [308]. There have been several improvements to the original scheme [308]. The as-synthesized nanocrystals, present in the A1 phase, transform into the $L1_0$ phase upon annealing at 560°C. The transition temperature can be varied by introducing other metal ions. Thus, $[Fe_{49}Pt_{51}]_{88}$ nanocrystals have been made by introduction of silver acetylacetonate in the reaction mixture [309,310]. The A1 to $L1_0$ transition temperature is lowered to 400°C by the introduction of Ag ions. Cobalt and copper ions introduced in Fe–Pt nanocrystals by the cobalt acetylacetonate or copper(II) bis(2,2,6,6-tetramethyl-3,5-heptanedionate) resulted in an increase in the A1 to $L1_0$ transition temperature [311,312]. The successful synthesis of Fe–Pt nanocrystals using a combination of reduction and thermal decomposition to successfully generate homogeneous alloy nanocrystals has sparked a flurry of activity. Thus, CoPt, FePd, and $CoPt_3$ have all been obtained [313]. Manganese–platinum alloy nanocrystals have been obtained by using platinum acetyacetonate and $Mn_2(CO)_{10}$ using a combination of reduction and thermal decomposition brought about using 1,2-tetradecanediol in a dioctyl ether, oleic acid/amine medium [314]. Ni–Fe alloy nanocrystals have been obtained using iron pentacarbonyl and $Ni(C_8H_{12})_2$ [315]. Sm–Co alloy nanocrystals have been prepared by thermolysis of Cobalt carbonyl and Samarium acetylacetonate in dioctyl ether with oleic acid [316]. Similarly, $SmCo_5$ nanocrystals have been prepared using a diol and oleic acid/amine [317].

Nanocrystals of metal oxides are prepared by controlled oxidation of the corresponding metal particles. Bentzon and coworkers prepared iron oxide nanoparticles, by exposing Fe nanoparticles, obtained by decomposing $Fe(CO)_5$ in decalin in the presence of oleic acid, to air for several weeks [318]. By adapting a similar route, but bringing about the oxidation by using $(CH_2)_2NO$, Hyeon and coworkers have obtained γ-Fe_2O_3 nanoparticles [319]. The same group has extended this method to produce cobalt ferrite nanocrystals [320].

Alivisatos and coworkers [321] prepared metal oxide nanoparticles by the decomposition of cupferron complexes in trioctylamine solution containing octylamine. Cupferron (*N*-nitrosophenylhydroxylamine) is a versatile ligand that forms complexes with several transition metal ions. By using this method, nanoparticles of γ-Fe_2O_3 [321], Cu_2O [321], Mn_3O_4 [321], Fe_3O_4 [322], and Co_3O_4 [322] have been made. The method has been used to the synthesis of nanocrystals of MnO, CoO, NiO, CuO, and ZnO [44, 45]. Caprylate capped γ-Fe_2O_3 nanocrystals are obtained by the thermolysis of Fe(III) hydroxide caprylate in tetralin [323]. Decomposition of the acetylacetonates in hexadecylamine yields amine-capped nanoparticles of Fe, Mn, Co, and Ni oxides [324].

Nanocrystals of perovskite oxides have been obtained by the thermal decomposition of MOCVD reagents (alkoxides such as $BaTi(O_2CC_7H_{15})[OCH(CH_3)_2]_5$) in diphenylether containing oleic acid and oxidizing the product with H_2O_2 [325]. Nanoparticles of $BaTiO_3$ and $PbTiO_3$ in addition to TiO_2 are obtained by this method.

Semiconductor Nanocrystals

The synthesis of some of the semiconducting metal chalcogenide nanocrystals was discussed in an earlier section. Murray and coworkers [326] in a pathbreaking paper described a method for synthesizing CdSe nanocrystals by reacting a metal alkyl (dimethylcadmium) with TOPSe (tri-*n*-octylphosphine selenide) in TOP (tri-*n*-octylphosphine), a coordinating solvent that also acts as the capping agent (see Fig. 2.14). This method readily yields CdS and CdTe nanocrystals as well. The reaction scheme of Murray succeeds to some extent in separating the nucleation and growth steps. When the chalcogen source is injected into the hot solution, explosive nucleation occurs, accompanied by a fall in temperature. Further growth occurs by maintaining the reagents around 100 K or lower, the final size depending on the growth temperature. The scheme of Murray has proved to be extraordinarily popular and has received extensive attention from numerous groups. Alivisatos and coworkers [327] have produced such nanocrystals by employing tri-butylphosphine at higher temperatures.

Peng et al. [328–331] have proposed the use of greener Cd sources such as cadmium oxide, carbonate or acetate instead of the pyrophoric

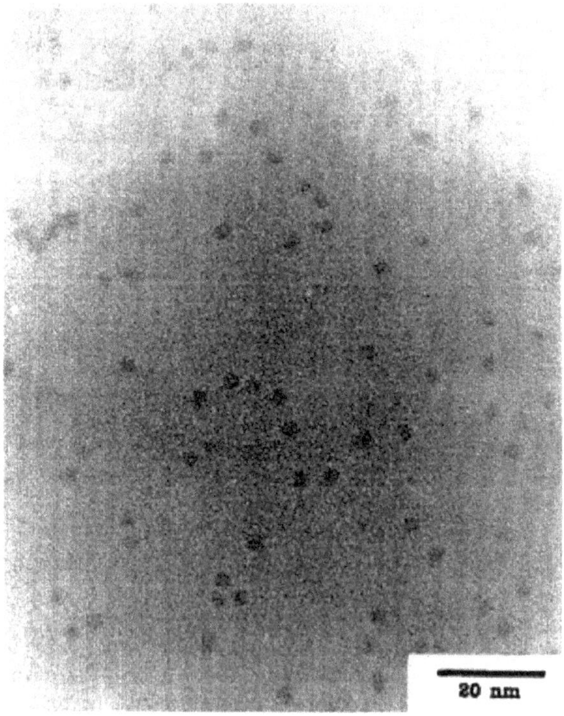

Fig. 2.14. TEM image of CdSe nanocrystals produced by the Murray method. The nanocrystals are elongated along one axis (reproduced with permission from [326])

dimethylcadmium. These workers suggest that the size distribution of the nanocrystals is improved by the use of hexadecylamine, a long-chain phosphonic acid or a carboxylic acid. The method can be extended to make CdS nanoparticles by the use of TOPS (tri *n* octylphosphine sulphide) and hexyl or tetradecyl phosphonic acid in mixture with TOPO–TOP. Fischer and coworkers [332] prepared TOPO-capped CdSe nanocrystals starting from less dangerous Cd precursors, while Guyot-Sionnest and coworkers [333] used dimethylzinc and a mixture of TOPO and hexadecylamine as capping agents to synthesize ZnSe nanocrystals. PbSe nanocrystals are prepared by the thermolysis of lead oleate and TOPSe in diphenylether–TOP [334]. Hyeon and coworkers [335] have prepared a series of metal sulphide nanoparticles such as ZnS, PbS, and MnS by the thermolysis of metal–oleylamine complexes in the presence of S and oleylamine. Nanocrystals of Ni_3S_4 and $Cu_{1-x}S$ are prepared by adding elemental sulfur to metal precursors dissolved in dichlorobenzene or oleylamine at high temperatures [336].

CdSe and CdTe nanocrystals can be prepared without precursor injection [337]. The method involves refluxing the cadmium precursor with Se or Te in octadecene. CdSe nanocrystals are also obtained by using elemental selenium dispersed in octadecene without the use of trioctylphosphine [338]. CdSe nanocrystals can also be obtained by a microwave-assisted route [339]. High-temperature synthesis of CdSe nanocrystals in nanoliter-volume droplets flowing in a perfluorinated carrier fluid through a microfabricated reactor has been described [340]. This method could be useful for providing a precise control in chemical or biochemical reactions. The kinetics of the formation of CdSe nanocrystals in trioctylphoshine oxide and stearic acid has been studied by following the changes in the emission band [341]. ZnSe nanocrystals have been synthesized in a hot mixture of a long chain alkylamine and alkylphosphines [342]. PbTe nanocystals in the size range of 2.6–8.3 nm have been prepared with size distributions less than 7% by a rapid injection technique [343]. PbS nanocrystals have been obtained by reacting $PbCl_2$–oleylamine complex with S-oleylamine complex without the use of any solvents [344].

Single Molecule Precursors

Decomposition of single molecular precursors provides convenient and effective routes for the synthesis of semiconductor nanocrystals. In this method, a molecular complex consisting of both the metal and the chalcogen is thermally decomposed in a coordinating solvent. Initial attempts with dithio and diselno carbamates, $[Cd(E_2CNEt_2)]_2$ (E=S,Se) gave nanoparticles of CdS [345,346]. The nanoparticles were capped with TOPO, the reaction medium. CdSe nanocrystals have been produced starting with compounds of the form $RCd(S_2CNEt_2)$(R = neopentyl, methyl)(see Fig. 2.15). Diselenocarbamates with unsymmetrical R groups such as hexyl and methyl $(CdSe_2CNMeHex)_2$ were found to be good air stable precursors for CdSe nanoparticles [347]. ZnS and ZnSe nanoparticles were prepared starting from $[EtZn(E_2NEt_2)]_2$ [348]. In a similar manner, lead chalcogenide nanoparticles have also been prepared [349]. Nanocrystals of Cd, Hg, Mn, Pb, Cu, and Zn sulphides have been obtained by thermal decomposition of hexadecylxanthates in hexadecylamine and other solvents. A highlight of this work is the use of relatively

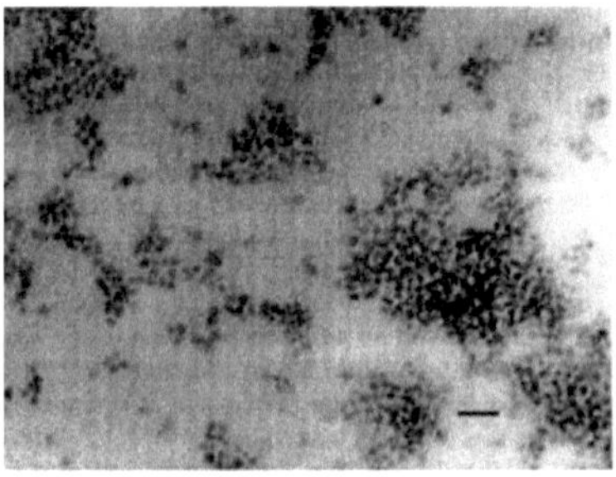

Fig. 2.15. TEM image of CdSe nanocrystals produced by the thermal decomposition of methyl diethyldiselenocarbamato cadmium(II). The scale bars correspond to 20 nm (reproduced with permission from [345])

low temperatures (50–150°C) and ambient conditions for synthesis of the nanocrystals [350]. Fluorescent CdSe nanocrystals have been prepared using air stable single source precursor cadmium imino-bis(diisopropylphosphine selenide) [351]. Water soluble luminescent CdS nanocrystals have been prepared by refluxing $[(2,2'\text{-bipyridine})Cd(SCOPh)_2]$ in aqueous solution [352]. Nanocrystals of EuS exhibiting quantum confinement have been synthesized by irradiating a solution of the dithiocarbamate $Na[Eu(S_2CNEt_2)_4]3.5H_2O$ in acetonitrile [353]. Following this initial report, a number of single source precursors have been used to synthesize EuS nanocrystals [354].

Nanoparticles of phosphides, nitrides, and arsenides have proven more difficult to obtain than the sulphides, selenides, and tellurides. An early procedure was the dehydrosilylation reaction proposed by the groups of Barron and Wells [355, 356]. Alivisatos and coworkers [357] adapted this method to prepare GaAs nanoparticles using $GaCl_3$ and $As(SiMe_3)_3$ in quinoline. Using a similar scheme, GeSb, InSb [358], InAs, and InP [359] nanoparticles have been obtained. The groups of Mícíc and Alivisatos modified the method of Murray and the Barron–Wells scheme to prepare InP, InAs, GaP, and $GaInP_2$ nanoparticles [360–364]. In a typical reaction, $InCl_3$ is complexed with TOPO/TOP and is reacted with a silylated pnictide such as $E(SiMe_3)_3$ (E = As,P) at 536 K, followed by growth at elevated temperatures for several days. Neither this scheme, nor the Barron–Wells scheme is readily extendable for the preparation of nitrides [365, 366].

Mícíc and coworkers [367] prepared GaN nanoparticles by thermolysis of polymeric $Ga(NH_{3/2})_n$ in trioctylamine. GaN nanoparticles have been prepared by controlled decomposition of gallium azide [368, 369]. Rao and coworkers [370] have established a single source route to nanostructures of the entire family of group 13 metal nitrides (GaN, AlN, InN). The precursors are the adducts of the metal chlorides and urea, namely $[Ga(H_2NCONH_2)_6]Cl_3$, $[Al(H_2NCONH_2)_6]Cl_3$ and $In(H_2NCONH_2)_3Cl_3$. They were decomposed under different conditions to yield nanocrystals. Hexagonal nanocrystals of GaN, AlN, and InN nanocrystals were obtained by refluxing the precursors in tri-n-octylamine. Remarkably, the nanocrystals of nitrides with the cubic structure could also be obtained employing solvothermal conditions.

Synthetic strategies that do not adopt dehalosilylation have also been reported [371]. Thus, Green and O'Brien prepared InAs nanoparticles by reacting $InCl_3$ with $As(NMe_2)_3$ in 4-ethylpyridine [372]. The dimer $[^tBu_2AsInEt_2]_2$ has been synthesized and used as a single-source organometallic precursor to grow InAs nanoparticles [373]. Efforts to prepare such nanoparticles by using single molecular precursors have met with some success. Douglas and Theopold [374] prepared InP nanoparticles by thermolysis of $[Cp^*(Cl)In(\mu\text{-}P(SiMe_3)_2)]_2$. O'Brien and coworkers [375–377] have developed a single molecular precursor route to synthesize InP and GaP nanocrystals using diorganophosphides -$M(PBu^t_2)_3$ (M=Gs, In). The nanocrystals are produced by refluxing the above in 4-ethylpyridine. This method has also been adapted to synthesize Cd_3P_2 using $[MeCdP(Bu^t)_2]_3$.

Korgel and coworkers [378–380] have reported a method to prepare Si nanocrystals at high temperatures in supercritical hexane and cyclohexane. The Si nanoparticles were capped with octanol, octene, and octanethiol.

2.2.7 Sonochemical Routes

The effect of ultrasound on a colloidal system has been known for sometime although its use for the preparation of nanosized matter is of relatively recent origin. Numerous methods have been discussed in the literature for the sonochemical synthesis of nanosized particles. However, not all the nanosized particles so obtained have been dispersed in a liquid medium. Progress in sonochemical synthesis made over the last two decades is illustrated by the set of examples discussed later.

In order to carry out sonochemical reactions, a mix of reagents dissolved in a solvent is subjected to ultrasound radiation (20 kHz–10 MHz). Acoustic cavitation leads to the creation, growth, and collapse of bubbles in the liquid medium. The creation of bubbles is due to the suspended particulate matter and impurities in the solvent. The growth of a bubble by expansion leads to the creation of a vacuum that induces the diffusion of volatile reagents into the bubble. The growth step is followed by the collapse of the bubble which takes places rapidly accompanied by a temperature change of 5,000–25,000 K in about a nanosecond. Collapse of the bubble triggers the decomposition of the matter within the bubble. The rapid cooling rate often hinders crystallization, and amorphous products are usually obtained. The collapse of the bubble does not signal the end of the reaction. The collapse is frequently accompanied by the formation of free radicals that cause further reactions. A few of the sonochemical reactions are, in fact, mediated by free radicals.

Baigent and Muller [381] reported sonochemical method to synthesize colloidal Au as early as 1980. This was followed by Henglein and coworkers [382]. The workers carried out sonochemical reduction of $HAuCl_4$ in an aqueous medium. This reaction has since achieved considerable attention [383–387]. By the use of PVP, the stability of the colloids is improved. The introduction of alcohols in the reducing mixture, enhances the rate of formation of the Au particles [383–385]. At high alcohol concentrations, smaller nanoparticles are formed. A mechanism based on the ability of alcohols to scavenge the H and the OH radicals has been proposed to account for these observations. Increasing the hydrophobicity of alcohols reduces the size of the nanoparticles due to the increasing ability of the hydrophobic alcohols to cap the produced nanoparticles. This method has been extended to prepare Pt nanoparticles, starting from chloroplatinic acid solutions containing ethyl alcohol [388]. Maeda and coworkers [389] used sodium dodecyl sulphate (SDS) as the stabilizing agent to obtain Pt nanoparticles of 2.6 nm diameter. The sonochemical reduction of Pd and Pt salts starting from K_2PdCl_4 and H_2PtCl_6 have been carried out under inert atmospheres [390]. The role of surfactants has been studied [391–393]. Pd nanoparticles in the size range 5–100 nm are obtained from aqueous solutions of $PdCl_2$ in the presence of stabilizing agents such as SDS, PVP, and poly(oxy-ethylene sorbitan monolaurate) [393]. The initial report of Henglein and coworkers [382] included a method for the synthesis of Ag nanoparticles. Maeda and coworkers [394] have obtained Ag nanocrystals by the sonochemical reduction of aqueous $AgClO_4$ and $AgNO_3$ in the presence of surfactants such as SDS. The average size of the particles obtained by the above methods are between 10 and 20 nm. Gedanken and coworkers [395] have prepared 20 nm sized Ag nanoparticles by the sonochemical reduction of $AgNO_3$ without the aid of capping agents. Metallic Cu nanoparticles have been obtained by the ultrasonic irradiation of copperhydrazinecarboxylate in water [396]. The reaction is believed to be mediated by hydrogen radicals.

Sonochemical treatment of volatile organometallic precursors dissolved in less-volatile solvents helps in the selective decomposition of the precursors. This method was first used by Suslick and coworkers to prepare colloidal Fe, by the decomposition of $Fe(CO)_5$ in octanol [397]. With PVP as the capping agent, the particle diameters obtained were in the 2–8 nm range, while oleic acid yielded monodispersed nanoparticles with an average diameter of 8 nm. Gadenken and coworkers [398–400] extended this method to prepare Co, Ni and Fe_2O_3 nanoparticles. Co nanoparticles stabilized by oleic acid were obtained in decalin, while oleic acid capped Fe_2O_3 nanoparticles were obtained in hexadecane. Ni particles could be prepared starting with $Ni(CO)_4$ in decane.

Organosols of Pd have also been obtained by the sonochemical reduction of $Pd(O_2CCH_3)_2$ and myristyltrimethylammonium bromide in THF or methanol [401]. Gadenken and coworkers [402] have obtained luminescent Si nanoparticles by the sonochemical reduction of tetraethoxyorthosilicate (TEOS) with a colloidal solution of Na in toluene. Si nanocrystals have been prepared by ultrasonically dispersing porous Si [403,404]. A variety of solvents were found suitable to prepare such dispersions.

Grieser and coworkers [405] obtained CdS nanoparticles by sonochemical means, starting with mixtures of Cd salts, water soluble thiols such as 2-mercaptopropanoic acid and sodium polyphosphate. The necessary sulphide ions were generated by the reaction of thiols with hydrogen radicals. Polyphosphate-stabilized nanoparticles were so obtained. By a similar procedure, CuS and PbS nanoparticles have also been prepared. By the ultrasonic irradiation of a mixture of metal salts and selenium in ethylene diamine, Li et al. [406] have obtained Ag, Cu, and Pb selenides. A sonochemical method for the synthesis of hollow MoS_2 nanocrystals has been described [407]. $BaFe_{12}O_{19}$ nanoparticles could be prepared by high-intensity sonication of a solution containing $Fe(CO)_5$ and barium ethylhexonate in decane, followed by calcination of the resulting powder at 900 K. After calcination, the particles were dispersible without the use of surfactants [408]. InP nanoparticles have been prepared by the ultrasonic irradiation of $InCl_3$, KBH_4, and yellow P in benzene and ethanol [409].

2.2.8 Micelles and Microemulsions

Reverse or inverted micelles formed by the dispersion of water in oil, stabilized by surfactants are useful templates to synthesize nanoscale particles of metals, semiconductors, and oxides [410–413]. This method relies on the ability of surfactants in the shape of truncated cones (like cork stopper), to trap spherical droplets of water in the oil medium, and thereby forming micelles. A micelle is designated inverse or reverse when the hydrophilic end of the surfactant points inward rather than outward as in a normal micelle. The dimensions of the water droplet can be suitably altered by changing the concentration of the surfactant and water. An inverted micelle is also called a microemulsion if larger water droplets are present. Appreciable diffusion of the matter dispersed in the water phase occurs in inverted micelles, caused by collision of water droplets. In order to synthesize nanocrystals, one mixes micelles made of aqueous metal ions with micelles consisting of an appropriate reagent that can bring about reduction, hydrolysis, or precipitation. Nanocrystals are formed upon mixing the micelles since crystallization is limited to within the confines of the water droplets. Reactions which yield metal oxides or chalcogenides in the bulk phase are carried out using inverted micelles to produce nanoparticles. Precipitation of the micelle with a polar solvent helps to recover the nanoparticles.

The first reported synthesis of nanoparticles using microemulsions was by Boutonnet and coworkers [414] who synthesized Pt, Pd, Rh, and Ir nanoparticles by reducing the corresponding metal salts with hydrogen or hydrazine in microemulsions. Pileni et al. [415] synthesized CdS nanocrystals by using the iso-octane-sodium bis(2-ethylhexyl)sulfo succinate (commonly called AOT or Aerosol-OT)-water system. The report was quickly followed by others [416] and the iso-octane-AOT-water system has emerged as a popular choice for the templated synthesis of nanoparticles. The iso-octane water system has been used to prepare nanoparticles of Cu, Ag, Ag_2S, AgI, AgI, etc. [411–413]. The relation between the relative concentrations of the surfactant (AOT) and water ($\omega = [H_2O]/[AOT]$) and the water pool radius (R, in nanometers) has been worked out (see Fig. 2.16). For values of ω upto 15, the following relation holds:

$$R = 0.15w \tag{2.4}$$

Cationic surfactants such as didodecyldimethylammoniumbromide (DDAB), neutral surfactants such as the family of ethoxylated alcohols of the type – CH_3–$(CH_2)_{i-1}$–$(CH_2CH_2O)_jOH$ (abbreviated as C_iE_j) –C_6E_5, $C_{10}E_8$ have also been used to prepare micelles. Zhang and coworkers [417] have synthesized $MnFe_2O_4$ particles using an inverted micelle template. One may consider that control over the size of the nanocrystals is achieved by altering ω. However, the dependence is often nonlinear, probably because the structure of water in the pool is different from that in the bulk [411–413]. Microemulsions made of water in supercritical fluids such as CO_2 have been used to synthesize nanoparticles of Ag, Ir, and Pt [418]. The ease of inverted micelle synthesis

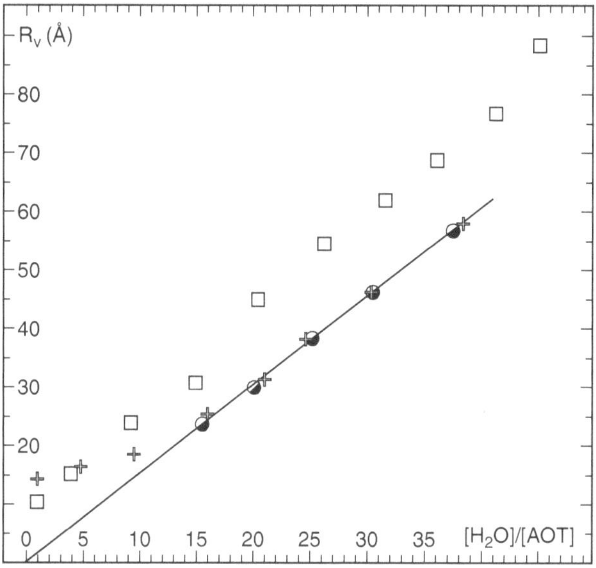

Fig. 2.16. Variation of the radius of the water pool in the iso-octane-AOT-water inverted micellar system. The water pool radius was estimated using Small angle X-ray scattering (*open squares*), Small angle neutron scattering (*crosses*), kinetic studies using hydrated electron probe (*open circles*) (reproduced with permission from [411])

has made it a potential method for industrial production of nanoparticles. Recent advances have tended to focus on methods for efficient recovery of nanoparticles. For example, supercritical CO_2 is proposed as a good medium for recovering nanocrystals [419].

Surfactants form inverted micelles in oil medium without the use of water. Dry powders of salts can, therefore, be dispersed in the surfactant solution and reduced with reducing agents such as $LiBH_4$ and Na in oil. Extremely reactive and even pyrophoric reducing agents can be used since water is absent in the system. Wilcoxon and coworkers [420–422] have made use of this method to synthesize Au, Si, and Ge nanocrystals. Si and Ge nanocrystals were obtained by dispersion of the halides in an inverted micelle followed by reduction with $LiAlH_4$ in THF.

Pileni and coworkers [423, 424] have pioneered the use of oil in water micelles to prepare particles of $CoFe_2O_4$, γ-Fe_2O_3, and Fe_3O_4. The basic reaction involving hydrolysis is now templated by a micellar droplet. The reactants are introduced in the form of a salt of a surfactant such as SDS. Thus, by adding CH_3NH_3OH to a micelle made of calculated quantities of $Fe(SDS)_2$ and $Co(SDS)_2$, nanoparticles of $CoFe_2O_4$ are obtained. By increasing the concentration of metal salts, the size of the nanoparticles can be increased. This method has been extended for the synthesis of Cu nanoparticles [425].

Sastry and coworkers [426] have made use of aqueous foams as templates to form platelets and particles of Au. A foam is generated in a column by the use of the surfactant CTAB (cetylthimethylammonium bromide) complexed with $HAuCl_4$. The Au ions dispersed over the foam are reduced by a gaseous reductant such as hydrazine vapors. It is supposed that the platelets are produced by reduction of the CTAB–Au complex molecules at the bubble walls, while the same complex molecules at the interbubble voids provide the isotropic space required for the generation of spherical particles.

2.2.9 The Liquid–Liquid Interface

Rao and coworkers [427–431] have used reactions taking place at the interface of two liquids such as toluene and water to produce nanocrystals and films of metals, semiconductors, and oxides. In this method, a suitable organic derivative of the metal taken in the organic layer reacts at the interface with the appropriate reagent present in the aqueous layer to yield the desired product. For example, by reacting $Au(PPh_3)Cl$ in toluene with THPC in water, nanocrystals of Au can be obtained at the interface of two liquids. This method has been extended to prepare nanocrystals of Ag and Pd, Au–Ag alloys, semiconducting sulphides such as CdS, ZnS, and CoS, and oxides such as Fe_2O_3 and CuO (see Fig. 2.17) [428–431]. By an appropriate choice of the reaction parameters, it has been possible to obtain isolated nanocrystals with narrow size distribution or well-formed films of the nanocrystals. By varying parameters such as the reaction temperature, and the reactant concentrations, the size of the nanocrystals and the coverage of the films can be modified. Thus, a change in the reaction temperature from 298 to 348 K, increases the size of Au nanocrystals from 7 to 16 nm (see Fig. 2.18). Starting with a mixture of metal precursors, it has been possible using this method to prepare Au–Ag alloy nanocrystalline films of varying compositions [432].

2.2.10 Biological Methods

Of the templates and systems used for the synthesis of nanocrystals, microbes offer an interesting possibility. The innards of a microorganism can be a tiny

Fig. 2.17. Nanocrystals of: (**a**) Au, (**b**) CdS, and (**c**) γ–Fe_2O_3 formed at the toluene-water interface (reproduced with permission from [427])

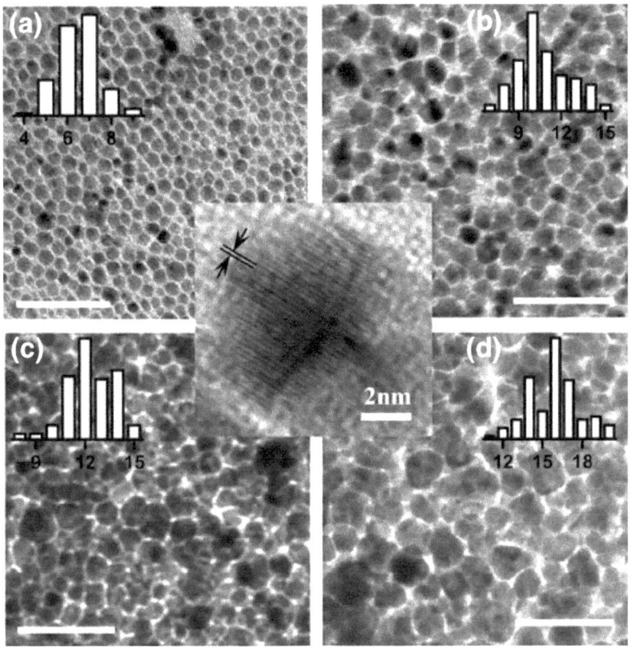

Fig. 2.18. Transmission electron micrographs of ultrathin nanocrystalline Au films prepared at the liquid–liquid interface at (**a**) 303 K (**b**) 318 K (**c**) 333 K, and (**d**) 348 K. The histograms of particle size distribution are also shown. The scale bars correspond to 50 nm. A high-resolution image of an individual particle is shown in the center (reproduced with permission from [431])

reactor as well as a container. Elementary reactions such as reduction are generally mediated by enzymes. Synthesis can therefore be carried out by simply incubating a solution of metal ions in the right microbial culture.

The ability of microbes to accumulate inorganic particles such as Au [433–435], CdS [436–438], ZnS [439], and magnetite [440] is well documented in the literature. It is also known that microorganisms put nanoscale particles to use as UV shields (CdS particles) and direction indicators (magnetite). The possibility of harnessing microorganisms for the synthesis of nanocrystals was realized only recently [441, 442].

Nair and Pradeep [443] have utilized Lactobacillus present in yogurt to synthesize Au, Ag, and Au–Ag alloy nanoparticles. The nanoparticles thus produced were in the size range of 15–500 nm. Joerger and coworkers [444–446] have synthesized Ag nanoparticles in the size range 2–200 nm by using *Pseudomonas Stutzeri. Klebsiella aerogenes* has been used to synthesize CdS nanoparticles in the size range 20–200 nm [447]. Roh and coworkers [448] have substituted metal ions such as Co, Cr, and Ni in magnetite nanocrystals synthesized using the iron-reducing bacteria *Thermoanaerobacter ethanolium*. Enzymes act as catalysts for the growth of metal nanoparticles. Enzyme mediated growth of metallic nanoparticles can be exploited for various purposes in biology involving dip pen lithography [449].

Apart from bacteria, yeast, and fungi have been used to obtain nanoparticles. Yeasts, *Candida glabrata* and *Schizosaccharomyces pombe* have been shown to yield CdS nanoparticles [437,438]. Kowshik et al. [450] have identified the ability of yeast *Torulopsis* sp. to produce nanoscale PbS nanoparticles. Sastry and coworkers [451–454] have identified two fungi species, *Fusarium oxysporum* and *Verticillium* sp. to produce Au and Ag nanoparticles. *Fusarium oxysporum* also reduces $CdSO_4$ to CdS to yield CdS nanoparticles [455]. CdS nanoparticles have been produced in the extracellular space. Highly luminescent, water-soluble and biocompatible CdTe nanocrystals have been prepared by using glutathione as a stabilizer. Quantitum yields in excess of 60% have been observed with these nanoparticles [456].

A novel nature of such biological synthetic schemes is that they produce nanoparticles at room temperature in aqueous medium, although poor size and morphology control also appear to be characteristic of these routes. Besides control over size and morphology, identification of the active biological ingredient that brings about the reaction remains unknown.

2.2.11 Hybrid Methods

Hybrid methods of preparation seek to combine both physical and chemical methods to produce nanoparticles. In these methods, the ability of physical methods to produce monodisperse nanocrystals is coupled with the ease of chemical synthesis to overcome the barrier of low throughput of physical processes. The method that combines the best of both the physical and chemical methods are that due to Andres and coworkers [457]. They were able to direct a mass-selected cluster beam onto a mixture of toluene containing thiol. The monodisperse nanocrystals in the beam get solvated and are capped by alkane thiols. This apparatus, christened multiple expansion cluster sources (MECS), is capable of producing truly monodisperse Au nanocrystals in solution form. MECS generated clusters have played a key role in enhancing our understanding of nanoscalar phenomena.

2.2.12 Solvated Metal Atom Dispersion (SMAD)

The SMAD method relies on trapping metal atoms in a cold matrix formed by a condensed liquid, followed by a growth step that occurs when the matrix is warmed to room temperature. By controlling factors such as solvent polarity and the rate of heating, the size of the nanocrystals can be tuned. The reaction is carried out in a vacuum chamber filled with liquid vapors. The walls of the chamber are cooled to liquid nitrogen temperature and the metal atoms are introduced by thermally evaporating a metal chunk. The atoms upon evaporation get embedded in the solvent matrix condensed on the walls. Klabunde and coworkers [458] have adapted this method to synthesize Au nanocrystals in gram scale. They have modified the SMAD scheme to include alkanethiol in the cold matrix and obtained monodisperse nanocrystals following a size fractionation method.

Laser Synthesis

The plume generated by a laser upon impacting a metal surface is known to consist of small atomic condensates. Cotton and coworkers [459] carried out early work in trying to disperse the plumes in the form of nanoparticles in solution. In recent times, Kondow and coworkers [460–463] have developed this method to synthesize Au, Ag, and Pt nanoparticles. To prepare Au nanoparticles, an Au plate dipped in water containing SDS was irradiated with laser light of wavelength 1,064 nm. The size of the nanocrystals so obtained depends on the laser fluence and surfactant concentration. Further modification of the initial distribution is possible by continued laser irradiation that causes fragmentation of the nanoparticles.

2.2.13 Postsynthetic Size-Selective Processing

The search for a general postsynthetic method for fractionating nanocrystals on the basis of size has been motivated by the need to employ truly monodispersed nanocrystals to reduce the uncertainty in the conclusions drawn from experiments. Nanocrystals that are not redispersible in liquids are not readily amenable to fractionation. In fact, attempts to purify charge-stabilized nanocrystals (by dialysis, for example) invariably lead to the irreversible precipitation of the crystallites. The availability of nanocrystals that are sterically stabilized and are hence redispersible makes it possible to derive schemes for postsynthetic size processing. Traditional methods of separation such as chromatography and centrifugation cannot be adapted for nanocrystalline systems. Centrifugal separation, though feasible in principle, is somewhat cumbersome.

The method of choice for postsynthetic processing is size-selective precipitation. In this method, two miscible liquids, one of which is capable of dissolving nanoparticles and the other that causes precipitation, are employed. When dispersed in such a mixture, the nanocrystals slowly precipitate, starting with the largest size. By collecting the supernatants or the precipitates, the nanocrystals can be fractionated on the basis of size. The composition of the solvent mixture determines the rate of precipitation and can be suitably altered to obtain the desired fractions. Other methods for size fractionation include digestive ripening and spectral decomposition. The digestive ripening scheme involves holding the particles in reflux for prolonged periods to obtain monodisperse nanocrystals. The spectral decomposition scheme exploits the photochemical decomposition of nanocrystals and is usually applicable to semiconductor nanocrystals. Since there is a direct relation between size and the absorbance of semiconductor nanocrystals, subjecting a sol to intense irradiation at an off-maximum position results in the decomposition of nanoparticles that have a large absorption cross-section at that particular wavelength. These particles are usually larger or smaller than the desired size. The last two schemes are adaptable to charge-stabilized nanocrystals as well since they involve processing in the liquid state.

2.3 Nanocrystals of Different Shapes

The ability to grow nanocrystals of controlled shapes adds an important dimension to the tuning of nanoscalar properties. In addition to size-dependent properties, nanomaterials possess shape-dependent properties. For example, how would the melting point vary with thickness if one were to make nanosheets of Au? The realization of the existence of shape-dependent properties provided an impetus to synthetic schemes to generate nanomaterials in shapes other than spheres. A number of solution-based methods have emerged to synthesize rods, prisms, cubes, and different polyhedral shapes of metals, semiconductors, and other materials. The capping agents that render spherical nanocrystals soluble are also generally able to solubilize other shapes. In the following sections, some of the solution-based schemes are examined.

2.3.1 Shape-Controlled Synthesis of Metal Nanocrystals

A number of synthetic schemes have been employed to obtain anisotropic growth of metal nanocrystals [464–470]. The schemes make use of templates or seeds to influence the growth process. The seed-mediated method has been adapted to produce nanorods [464,465], nanowires [466–468], and other shapes [469, 470] of Au and Ag in aqueous media. The method involves two steps. In the first step, small citrate capped Au or Ag nanocrystals are produced by borohydride reduction for use as seeds. In the second step, the particles are introduced into a solution containing the metal salt, CTAB (a structure-directing agent), and a mild reducing agent such as ascorbic acid. The use of a mild reducing agent is the key to achieve seed-mediated growth. Under the reaction conditions, ascorbic acid is not sufficiently powerful to reduce the salt on its own. In the presence of seeds, reduction mediated by the seed occurs producing nanorods [464, 465] and nanowires [466–468]. A TEM image of Au nanorods produced by this method is shown in Fig. 2.19. The nanorod and nanowire structures are directed by the micellar structures adopted by CTAB. The chain length of the structure-directing agent plays an important role in determining the aspect ratio of the rod-shaped particles [467]. The presence of a small quantity of organic solvents leads to the formation of needle shaped crystallites [468]. Addition of NaOH to the reaction mixture before reduction brings about dramatic changes in the product morphology. Hexagons, cubes, and branched structures have been produced by using NaOH and varying the experimental parameters (see Fig. 2.20) [469]. Nanoplates of Ag have also been prepared by adopting a similar procedure [470].

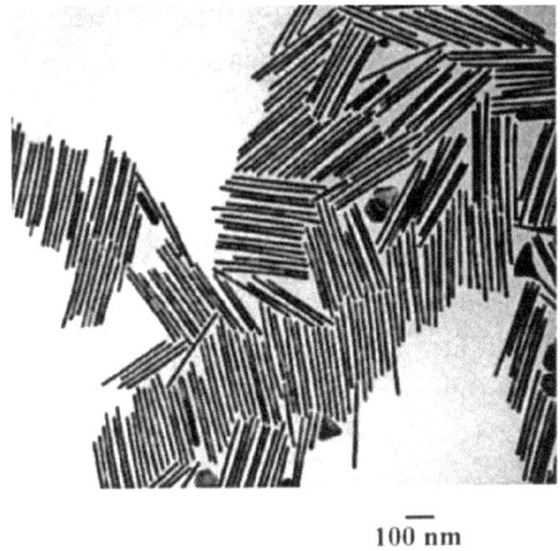

Fig. 2.19. TEM image of Au nanorods prepared by the seed mediated growth method (reproduced with permission from [466])

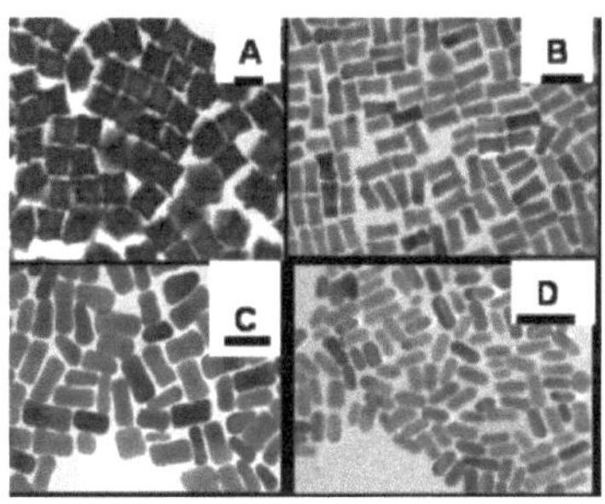

Fig. 2.20. TEM images showing cubic to rod-shaped gold particles produced with low concentrations of ascorbic acid in the presence of a small quantity of silver nitrate, by seed-mediated growth method. The concentration of CTAB increases from 1.6×10^{-2} M(A) to 9.5×10^{-2} M (B,C,D). [Au^{3+}] decreases from (B) to (C), whereas the seed concentration increases from C to D. Scale bar is 100 nm (reproduced with permission from [469])

By the use of NaOH, one can do away with both the surfactant and seeds and still obtain nanowires [471]. The mechanism operating in this reaction is not understood, but it is clearly not surfactant-directed. PVP-capped nanowires of Ag have been made by employing the polyol process with [472] and without seeds [473]. It is believed that PVP plays the role of the structure-directing agent. As with the seed mediated method, other shapes have also been synthesized by this method. By tuning the experimental conditions, Sun and Xia obtained nanocubes of Au and Ag [474]. Triangular Au nanoparticles or nanoprisms have been grown using lemon-grass extract [475]. Ag nanoparticles of a variety of morphologies are obtained by carrying out the reduction with silver binding peptides [476].

Mirkin and coworkers [477] have devised two different routes to nanoprisms of Ag. In the first method, Ag nanoprisms are produced by irradiating the mixture of the citrate and bis(p-sulphonatophenyl) phenylphosphine dihydrate dipotassium (BSPP) capped Ag nanoparticles with fluorescent lamp. By controlling the wavelength of irradiation, the nanoprisms can be induced to aggregate into large prisms in a controlled manner (see Fig. 2.21) [478]. In the second method, $AgNO_3$ is reduced with a mixture of borohydride and hydrogen peroxide [479]. The latter method has been extended to synthesize branched nanocrystals of Au (see Fig. 2.22) [480,481]. Xiao et al. [482] adapted a NAD(P)H-mediated growth in the presence of ascorbic acid to synthesize dipods, tripods, and tetrapods of Au. Other methods such as the inverted micelle method [483] and the DMF reduction method [484] can also be adapted to prepare nanoprisms or nanoplates of Ag.

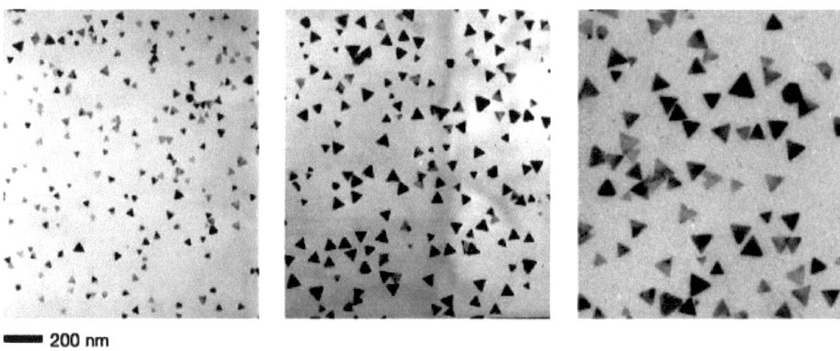

Fig. 2.21. Ag nanoprisms of different dimensions obtained by controlled irradiation of bis(p-sulphonatophenyl) phenylphosphine dihydrate dipotassium capped Ag nanoparticles (reproduced with permission from [478])

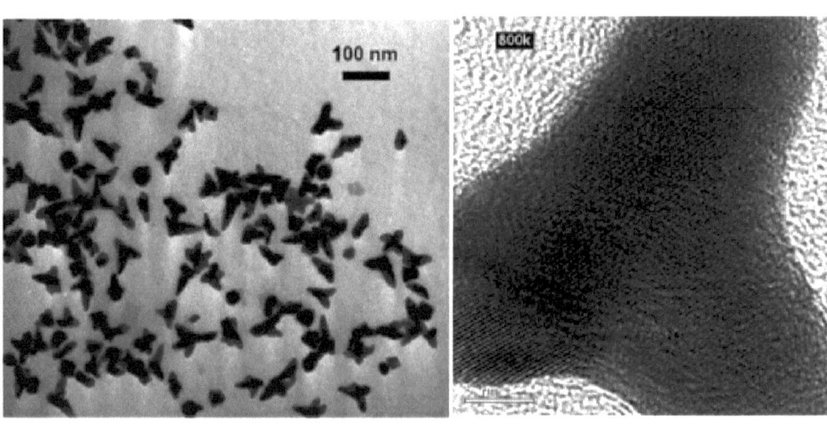

Fig. 2.22. Low and high-magnification TEM images of branched Au nanocrystals (reproduced with permission from [480])

An octlyamine/water bilayer template has been used to synthesize Ag nanoplates [485]. Au nanoplates have been produced by using aspartate as the reducing agent [486]. The methods employed for synthesizing Au and Ag nanostructures of different shapes are not readily extendable to other metallic systems. However, there are other methods. Hydrogen reduction of olefinic Co compounds has yielded Co nanorods [487]. Photochemical reduction of olefinic complexes of In and Sn in the presence of long-chain amines yields In and In_3Sn nanowires [488]. Pt nanocrystals in cubic, octahedral, and cuboctahedral shapes have been prepared by the polyol method [489]. Post-synthetic changes in the morphology of nanoparticles has also been achieved. Klabunde and coworkers [490] have obtained varied shapes of Au nanoparticles by changing the ratio of the competing reagents, more specifically, alkylthiols and tetralkylammonium salts. Pt nanocrystals of complex morphologies including dendrites, polypods, tetrapods, and tripods have been obtained by thermolytic reactions [307].

2.3.2 Shape-Controlled Synthesis of Semiconductor and Oxide Nanocrystals

Thermolysis is the method of choice for the shape-controlled growth of semiconductor nanocrystals. Anisotropic growth is achieved by introducing the precursor into a hot mixture in several small steps rather than in a single step. Thus, a high concentration of the precursor is maintained throughout the growth step. The first such report was on the nanocrystals of CdSe in the shape of rods, arrows, and tetrapods using the "green" sources (see Figs. 2.23 and 2.24) [328,491–493]. This scheme has been extended to synthesize CdS [494] and CdTe [495] nanocrystals of different shapes. Interestingly, ZnO tetrapods have been obtained by a physical vapor deposition process based on carbothermal reduction of ZnO [496,497].

Heterostructured tetrapods consisting of CdS, CdSe, and CdTe are prepared by a careful control of experimental conditions. Thus, arms of CdSe have been grown on CdS nanorods [498]. Rods and tetrapods have been obtained with other precursors as well [499]. By varying the quantity as well as the nature of the organic surfactant, the synthesis of hyperbranched CdSe and CdTe nanocrystals with fair degree of control over the length as well as branching has been reported [500]. CdS nanocrystals in the wurzite structure have been obtained in hexagonal or pyramidal geometries by adjusting the molar ratios of Cd and S precursors in the solution [501]. Thermal decomposition methods have been employed to synthesize nanorods and nanowires of FeP [502] and ZnS [503]. Iron oxide nanocrystals in cubic, star-shaped, and other shapes have been obtained by control of the thermolysis conditions [504]. In the hot injection method, the presence of acetate ions seems to favor the formation of star-shaped PbSe nanocrystals [505].

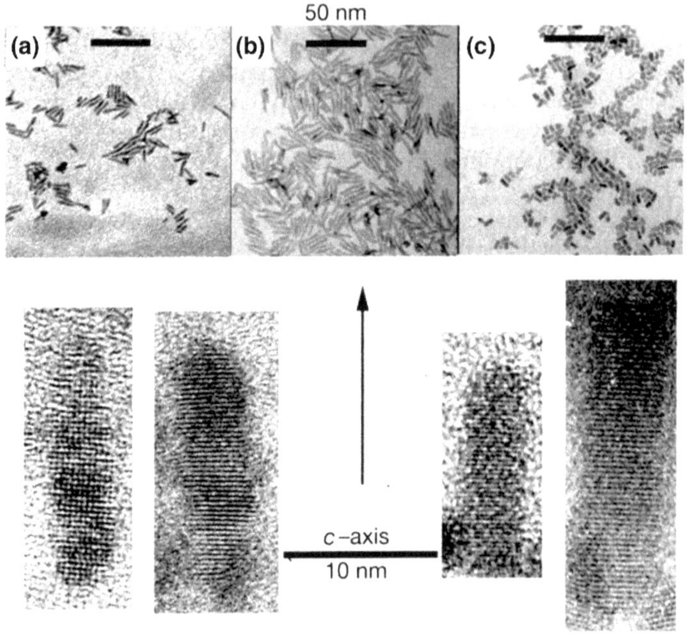

Fig. 2.23. (**a**–**c**) TEM images of nanorods of CdSe with different sizes and aspect ratios, high-resolution TEM images of four nanorods are shown (reproduced with permission from [492])

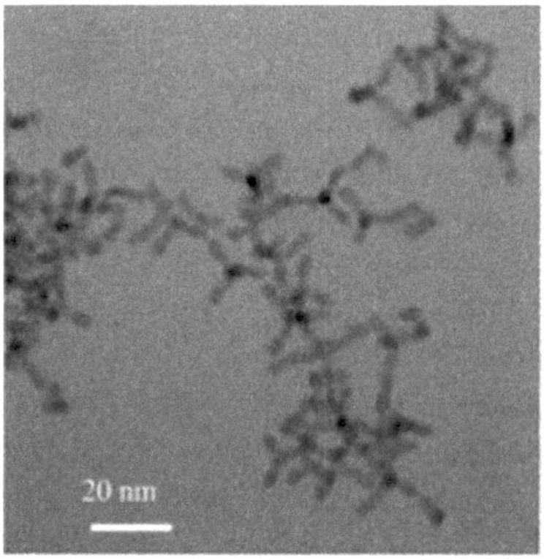

Fig. 2.24. TEM image of CdSe tetrapods (adapted with permission from [493])

A method closely related to thermolysis has been used by Korgel and coworkers [506–508] to synthesize nanocrystals in different shapes. In this method, long-chain alkanethiolates of metals such as Cu and Ni are mixed with fatty acids to obtain a waxy solid which is thermally decomposed. Thus, nanorods, nanodisks, or nanoprisms of Cu_2S [506, 507] and NiS [508] have been prepared by this method (see Fig. 2.25).

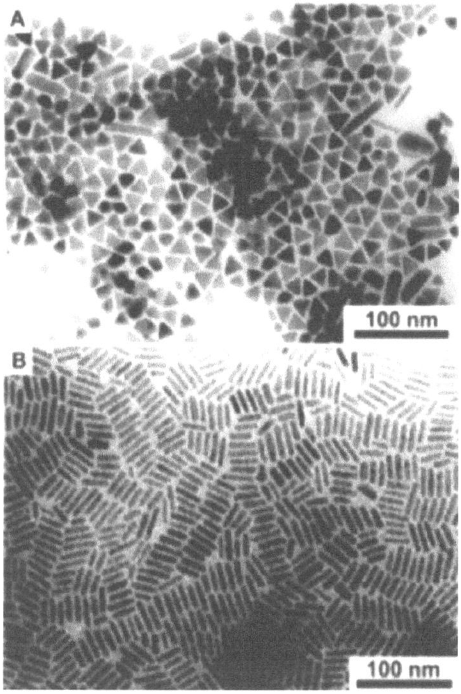

Fig. 2.25. TEM images showing nanoprisms and nanorods of NiS. The images correspond to different areas of the same sample (reproduced with permission from [508])

It is possible to rationally design templates for shape controlled synthesis of nanocrystals using the inverted micelle method. By varying w, cylindrical water pools of different dimensions can be obtained. The template can be used to grow nanowires. Elongated Cu nanoparticles [509], nanowires of $BaCO_3$ [510] and $BaSO_4$ [511] have been obtained by the use of such templates (see Fig. 2.26). The particles present in the water pool can undergo further changes. Thus, triangular CdS [512] as well as prismatic $BaCrO_4$ [513] have been prepared starting with an inverted micelle. Other ways to obtain shape control include controlled oxidation or reduction. Large tetrahedral Si nanocrystals have been obtained as exclusive products by careful control of the reducing conditions [514]. Nanorods and nanodisks of ZnO have been grown by a controlled room-temperature oxidation of dicyclohexylzinc [515]. ZnO nanoparticles with cone, hexagonal cone, and rod shapes are obtained by the nonhydrolytic ester elimination sol–gel reactions [516]. In this reaction, ZnO nanocrystals with various shapes were obtained by the reaction of zinc acetate with 1,12-dodecanediol in the presences different surfactants. Hexagonal pyramid-shaped ZnO nanoparticles have been obtained by thermolysis of a Zn-oleate complex [517].

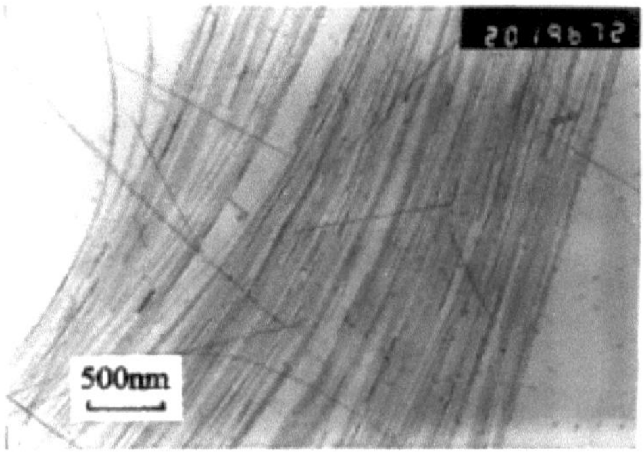

Fig. 2.26. TEM micrograph of $BaCO_3$ nanowires obtained by the inverted micelle method (reproduced with permission from [508])

2.4 Doping and Charge Injection

Intrinsic semiconductors need to be doped to modify their charge transport characteristics. The maximum doping normally used is of the order of 1 dopant atom per 10^5 atoms. Doping nanocrystals at such levels is unrealistic as their nuclearity is often below 10^5 atoms. Furthermore, nanocrystals possess a different band structure as well as excitonic character due to quantum confinement. Doping of nanocrystals is effected for entirely different reasons. Semiconductor nanocrystals are doped with few percent of impurities to create impurity centers that interact with the electrons and holes. A useful effect of this interaction is that the mid gap states arising from surface species can be shifted outside the gap region. Dopants do not affect the absorbtion spectra, however, the emission intensity is greatly increased. Doping is achieved by simply introducing the dopant in the reacting mixture. The most commonly investigated systems include Mn doped CdS [518–520], ZnS [521–523], CdSe [524], and ZnSe [525, 526]. Mn is a particularly good choice because proper Mn doping could result in the material turning into a dilute magnetic semiconductor with room temperature ferromagnetic behavior. Thermolysis is employed to produce doped nanocrystals as it yields monodispersed nanocrystals with good crystallinity. The biggest challenge facing doping studies is the tendency of crystallites to expel the dopant ions along with the other defects to the surface rendering doping ineffectual. Unfortunately, the rate and efficiency of this process increases at high temperatures. This process is also dependent on the host lattice. ZnSe nanocrystals have been successfully doped with Mn using thermolysis [525,526]. The same method yields surface-enriched rather than doped particles in other systems. A proper choice of precursors could help halt expulsion of dopants to the surface. Employing a precursor such as $Mn_2(\mu\text{-SeMe})_2(CO)_8$, CdSe nanocrystals have been doped successfully with Mn [524]. Recently, Norris and coworkers have suggested that the success of doping is related to the binding energy of the dopant ions to the exposed surface of the growing nanocrystal. High binding energies lead to successful adsorption and doping, while low binding energies mean that doping is not favored [527].

An attractive alternative to traditional doping is that suggested by Guyot-Sionnest [528]. It is possible to turn a semiconductor nanocrystal into n-type or p-type by directly injecting charge into the nanocrystal. This is accomplished electrochemically or by the use of strong reducing agents in solution. Thus, n-type CdSe and ZnS have been obtained through charge transfer brought about by reducing agents [529,530]. The success of the charge injection scheme can be discerned from the electronic absorption spectra. An effect associated with n-doping is the loss of fluorescence upon charge injection. The n-doped nanocrystals tend to oxidize under ambient conditions. One could envisage extending this scheme to other kinds of nanocrystals as well.

2.5 Tailoring the Ligand Shell

The ligand shell surrounding a nanocrystal provides the interface for interaction of the particles with the environment. The ligand shell can have pronounced influence on the properties of the nanocrystals. Solubilization of nanocrystals, a property associated with the ligand shell, is the basis for much of chemical and physical studies on nanocrystals. It is possible to transfer nanocrystals from aqueous to organic medium and vice versa by changing the ligands at the surface of the nanocrystal. The first such report concerned the transfer of 8 nm Ag nanoparticles from water to cyclohexane by the use of sodium oleate and NaCl [532]. Au_{55} nanocrystals were transferred from dichloromethane to water by displacing triphenylphosphine at the surface of the nanocrystal with triphenylphosphine sulfonate [533]. Such a transfer has also been accomplished with other ligands [534, 535]. Brown and coworkers [534] found that the phase transfer was accompanied by a growth of the nanocrystals.

A novel method of thiol-derivatizing hydrosols of noble metals has been reported by Sarathy et al. [536, 537]. The procedure involves the transfer of hydrosols to an organic medium using thiols, in the presence of a strong acid or a reducing agent. A simple modification of this technique is shown to be effective in the case of CdS nanocrystals as well [538]. The transfer of metal nanocrystals from organic to aqueous phase is possible by the use of dimethlyaminopyridine as a phase transfer reagent and capping agent [539, 540]. Metal, semiconductor, and oxide nanocrystals can be transferred from an organic to an aqueous phase by coating the nanocrystals with small polymer chain consisting of interdigitating chain molecules and a cross linker [541].

The possible use of fluorescent quantum dots as labels in biological environments has made it necessary to devise ways to transfer semiconductor nanocrystals from organic to aqueous medium. Furthermore, the new layer of a capping agent has to provide the means of interfacing the transferred nanocrystals to biological moieties. Phase transfer from an organic to an aqueous medium as well as the displacement of TOPO/TOP from the surface has been accomplished by silica coating [542] or by derivatizing the surface with bifunctional molecules such as mercaptoacetic acid [543], dithiothreitol [544], and lipoic acid [545]. The latter method results in less stable capping, but the nanocrystals so obtained could be bound directly to proteins and related structures via the pendant carboxyl moieties. Once bound to such large species, the stability of the nanocrystal is increased [546]. It is noticed that unidentate ligands desorb from the surfaces of nanocrystals over a period of time. Indeed, to obtain stable dispersions of nanocrystals, it is often necessary to add a small excess of the capping agent. Bidentate ligands bind more strongly to the surfaces of nanocrystals and are less prone to desorption. The highest stability is, however, provided by the silica shell. The silica shell can be further derivatized with silanes with phosphonate or ammonium head group to bestow a desired charge to the surface of the nanocrystal at a chosen pH [547]. The surface can be rendered neutral by the use of silane molecules with long chain hydrophilic chains such as poly(ethyleneglycol). Other approaches to make nanocrystals water soluble include displacement of TOPO/TOP with a designer peptide sequence that includes a hydrophilic sequence to solubilize the nanocrystal and an adhesive sequence to bind to target biomolecules. By employing different methods, surfaces of nanocrystals have been modified with polyelectrolytes [548] and exfoliated sheets of graphene oxide [549]. In some cases, surface modification of nanocrystals may not be required as they solubilize in both aqueous and organic media. One such example is Ru nanoparticles stabilized by oligoethyleneoxythiol [550].

Recently, fluorous chemistry has been used to solubilize and phase-transfer nanocrystals of metals (Au) and chalcogenides (CdS) as well as inorganic nanorods (ZnO) [551]. Here, a fluorous label (e.g. fluorinated alkane thiol) is attached to the nanostructures which are then readily extracted by a fluorocarbon from aquous or hydrocarbon solutions, due to fluorine–fluorine interaction. This method not only enables the purification of the nanostructures, but also helps one to study them in the most-nonpolar medium possible, the refractive index of fluorocarbon being around 1.2.

3

Programmed Assemblies

Just as atoms and molecules crystallize into different structures, capped nanocrystals spontaneously organize themselves into ordered aggregates. This statement is not entirely correct since crystallization of these artificial atoms, i.e., nanocrystals, is brought about by weak dispersive forces dependent on factors such as the size of the nanocrystal and the nature of the capping ligand. Some control over the process of self-assembly can be exercised by varying certain parameters. For example, by increasing the chain length of an alkanethiol, the interparticle separation distance between the metal nanocrystals can be increased. The realization that material properties can be tuned by continuously varying the separation distance has lent impetus to organizing nanocrystals into one, two, and three dimensions. In other words, ligated nanocrystals provide a means to realize a designer lattice, wherein one can control the lattice spacing as well as the lattice type. As stated by Jim Heath, one can thus tune covalency.

3.1 One-Dimensional Arrangements

Organization of nanocrystals in one dimension in the form of chains and strands has met with some success. Templates are commonly used to bring about such organization. The templates employed include host lattices such as pores in alumina, steps on crystalline surfaces as well as polymer chains. Hornayak and coworkers [552] used ordered channels of porous alumina as templates to obtain linear arrangements of Au nanocrystals. By varying the pore size, the diameter of the nanostructure can be varied. Different types of polymer chains are used as scaffolds to bring about one-dimensional organization of nanocrystals. Gold nanoparticles have been assembled into one-dimensional chains by the controlled ligand exchange of citrate ions with 2-mercaptoethanol [553]. Single and double stranded DNA are employed to obtain linear assemblies of Au and other nanocrystals. These assemblies are caused by the electrostatic interaction between the polyphosphate backbone

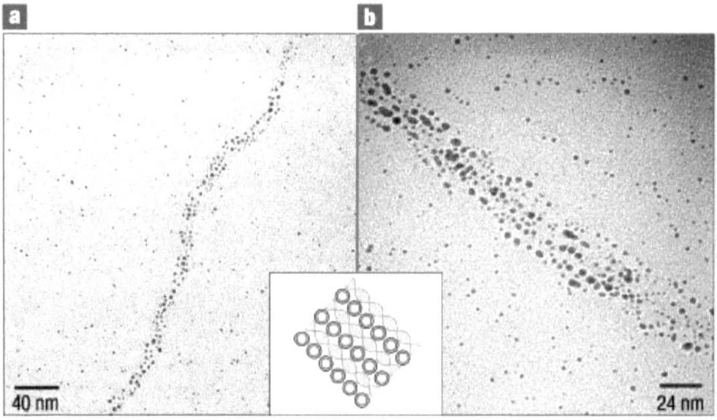

Fig. 3.1. TEM images of Au nanocrystals organized into (**a**) ribbons and (**b**) thicker strands on the DNA backbone. The inset illustrates schematically the relationship between the DNA chains and the Au nanocrystals (reproduced with permission from [555])

of the DNA and the cationic charges resident on the surface of Au nanocrystals [554, 555]. Thus, Warner and Hutchison [555] successfully prepared closely packed linear arrays, ribbons, and branched chains of Au nanocrystals as shown in Fig. 3.1. Alivisatos and coworkers [556] have covalently attached thiol modified single strands of DNA to Au nanocrystals. The ability of strands to bind to each other with specific sequence has been exploited to bring about linear and low-dimensional arrangements of nanocrystals. The binding affinity of single stranded DNA to Au nanoparticles is size dependent. Assemblies of 17 and 5 nm Au nanoparticles have been achieved using single stranded DNA without the need for hybridization [557].

ZnS nanocrystal aggregates have been organized into linear arrays by using a virus template [558]. Similarly, Pt nanocrystals in the form of ribbons have been obtained using a cholesteric liquid crystalline template [559]. By reducing copper ions bound to a fiber-forming lipid, Kogiso et al. [560] obtained a one-dimensional arrangement of Cu nanocrystals organized on the lipid backbone. Schmid and coworkers [561] have obtained a linear arrangement of Au_{55} nanocrystals by the use of a polymer fiber formed at the interface of two liquids. It is suggested that the tobacco mosaic virus tubules can serve as templates for the growth of 1D lattice of quantum dots [562]. Oku and Suganuma [563] have obtained a linear arrangement of Au nanocrystals (5 nm) at the step edges of a thin carbon film. The adhesive forces present at the step edge seem to guide the organization (see Fig. 3.2). The length of the organization depends on the length of the step edge.

Heath and coworkers [564] describe a rare example of one-dimensional organization in the absence of a template. They fabricated wires of Ag nanocrystals by compressing a dispersion of Ag nanocrystals (4.5 nm) in toluene at the air–water interface in a Langmuir–Blodgett trough. The wires were one

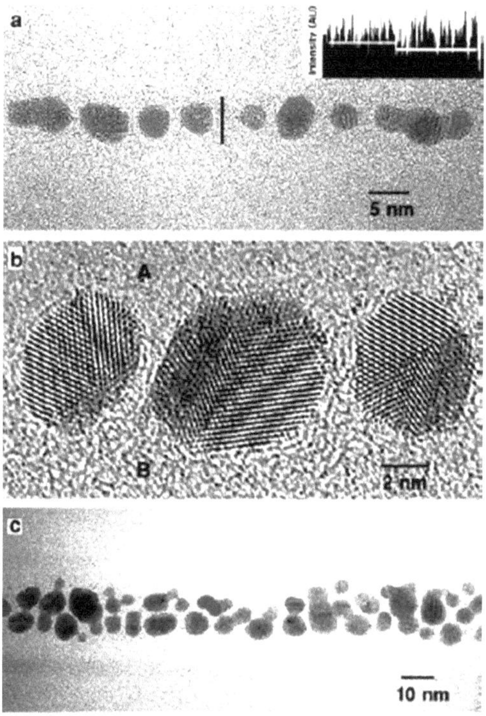

Fig. 3.2. (**a**) TEM image of Au nanocrystals organized in the form of a linear row along the stepedges of a carbon film. The location of the stepedge can be clearly seen in the intensity profile along a cross-section of the film shown in the inset. The profile corresponds to the area indicated by a *black line*. (**b**) High-resolution TEM image showing the lattice planes of Au nanocrystals. (**c**) Au nanocrystals organized on a wider step (reproduced with permission from [563])

nanocrystal thick, a few nanocrystals wide and extended in length from 20 to 300 nm. The interwire separation distance as well as the alignment of the wires could be controlled by varying the surface pressure of the film (see Fig. 3.3). Understanding the process involved in such an organization would help to make use of this process with other systems, and pave the way to template free one-dimensional organizations. Short gold nanorods have been linked to form chains by using a mercaptocarboxylic acid wherein the sulphur bonds to the metal and the carboxyls form hydrogen bonds [565]. The association was examined by electron microscopy and absorption spectroscopy.

Oriented attachment of nanocrystals can be used to make one-dimensional and other complex nanostructures. Thus, nanotubes and nanowires of II–VI semiconductors have been synthesized using surfactants [566]. The nanorods or nanotubes of CdS and other materials produced in this manner actually consist of nanocrystals. The synthesis of TiO_2 nanowires from nanoparticles has been reported [567].

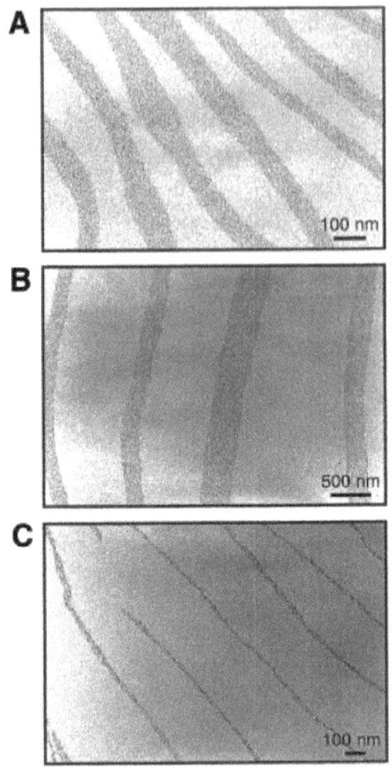

Fig. 3.3. TEM images of Ag nanocrystal wires of different thicknesses (**a**) Ag nanocrystals of diameter 3.4 nm deposited from hexane sol (**b**) Ag nanocrystals of diameter 3.4 nm deposited from heptane sol (**c**) Ag nanocrystals of diameter 4.4 nm deposited from heptane sol (reproduced with permission from [564])

3.2 Rings and Associated Arrangements

The templates used to prepare one-dimensional organizations cannot be extended to bring about organizations in higher dimensions. One would expect that it would be more difficult to obtain more complex structures such as rings and circles. However, thanks to the dynamics of solvent evaporation, it has been possible to prepare circular arrangements of nanocrystals without the aid of a template. The first observation of ring-like arrangement of nanocrystals was made by Heath and coworkers [568, 569]. They observed rings of diameters between 0.1 and 1.0 µm made up of thiol-covered Au or Ag nanocrystals in the size range of 2–6 nm (see Fig. 3.4). The phenomenon was explained as due to the formation of dry holes that expand during the evaporation of a wetting thin film [570]. Gedanken and coworkers [408] observed the formation of an olympic ring-like arrangement (intersecting rings) when freshly prepared bariumhexaferrite nanocrystals were deposited on a TEM grid (see Fig. 3.5). The formation of such rings is inconsistent with the model of dry hole expansion. The formation of intersecting rings had been attributed to an interplay between magnetic interactions among the particles and particle-surface interactions.

Pileni and coworkers [571,572] prepared rings as well as hexagons of nanocrystals of Ag, Cu, Co, CdS, and Ag_2S (see Fig. 3.6). The formation of these structures is attributed to Bénard-Marangoni instabilities caused by convection currents in an evaporating thin film. The convection currents could lead

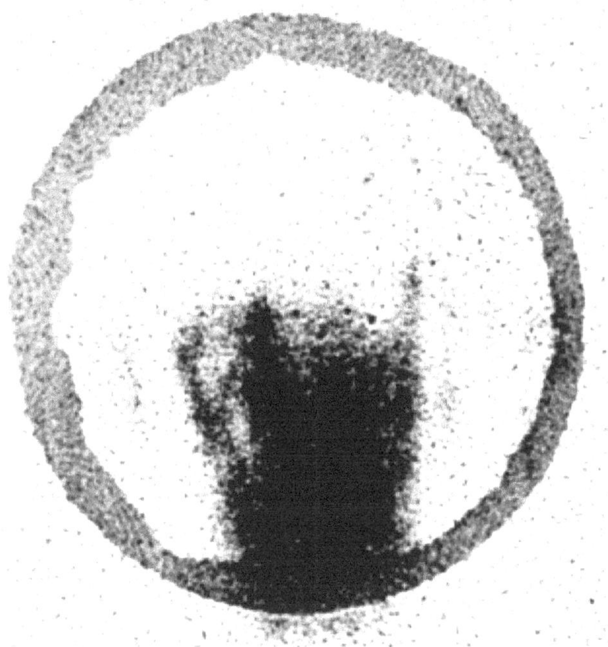

Fig. 3.4. TEM image of a ring made of nanocrystals observed upon evaporation of hexane from a dilute solution of 3 nm gold nanoparticles coated by dodecanethiol (reproduced with permission from [570])

to isolated as well as intersecting rings depending on the concentration of the nanocrystals in the evaporating droplet. The ring to hexagon transition has been studied in detail using Au and Ni nanocrystals with diameters of 3.5 and 4.0 nm, respectively [573]. By adopting a complex process involving phase separation in binary fluids, instabilities and evaporation of microdroplets, rings of magnetic $CoPt_3$ nanocrystals have been prepared [574]. Here, the diameter of the rings was in the range of a few microns and the ring walls were made up of a few nanocrystals. Rings of nanocrystals of much smaller dimensions have been prepared by other methods. Small rings of Co nanocrystals with diameters in the 50–100 nm range have been obtained by capping the nanocrystals with a specially designed ligand, C-undecylcalix[4]resorcinarene [291]. Ring-like structures of semiconductors are found to occur by the self-assembly of nanocrystals in aqueous media [575]. This organization is considered to be due to the self-assembly of the ligands. By tailoring the surface of a block copolymer film and depositing the nanoparticles atop these films, nanoscopic rings of Au nanocrystals with diameters in the range of 15–30 nm have been obtained [576].

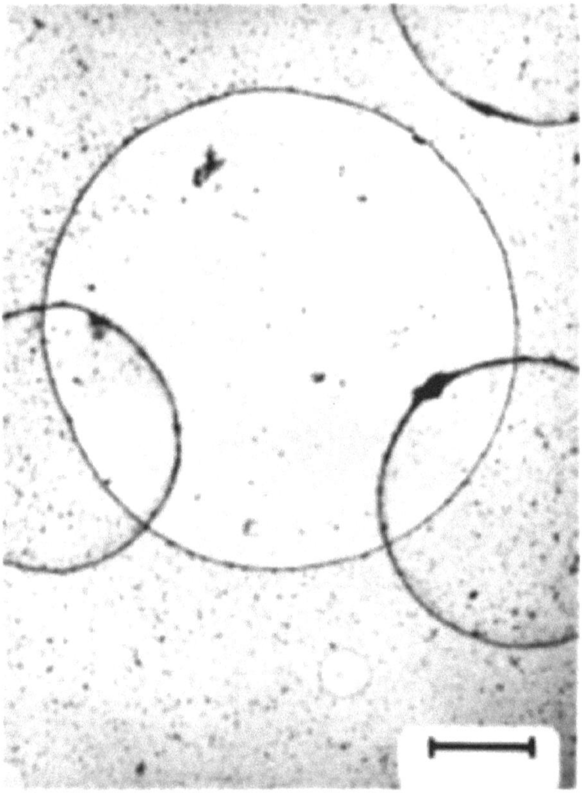

Fig. 3.5. TEM micrograph showing rings of barium hexaferrite nanocrystals. The rings intersect with each other forming olympic rings. The scale bar is 0.7 µm (reproduced with permission from [408])

Besides rings, spontaneous formation of nanoparticle strip patterns has been observed on dewetting a dilute film of polymer coated nanoparticles floating on a water surface [577]. Hybridization of branched DNA trimers and Au nanoparticle DNA conjugates have been employed to produce discrete self-assembled nanoparticle dendrimers [578]. Self-assembly of triangular and hexagonal CdS nanocrystals into complex structures such as rods and arrows has been observed [579]. Furthermore, self-assembly of CdSe nanoparticle–copolymer mixtures has been observed wherein the copolymers assemble into cylindrical domains that dictate the distribution of the nanoparticles [580].

3.3 Two-Dimensional Arrays

In contrast to the limited success with one-dimensional arrangements, two-dimensional lattices of nanocrystals have been obtained with great many variations. These arrays have generally been prepared by simply evaporating

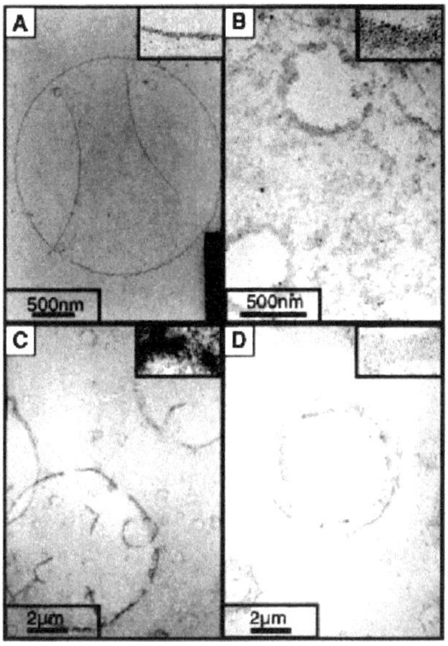

Fig. 3.6. TEM images of rings of (**a**) silver; (**b**) copper; (**c**) cobalt; and (**d**) silver sulfide nanocrystals. The rings were obtained by drying a dilute solution on the TEM grid (reproduced with permission from [571])

dispersions of suitably functionalized monodisperse nanocrystals on hydrophobic surfaces such as a carbon film. The cooperative assembly of nanocrystals is brought about by a weak force that is resultant of an attractive force between the nanocrystal cores in solution and a repulsive part arising from steric crowding of ligands as the nanocrystals are brought closer. To bring about large scale organization, a right balance of forces is essential. Thus, it is vital that the nanocrystals are monodisperse and the ligand molecule covers the nanocrystals. Ligands such as long chain thiols, amines, phosphines, or fatty acids have served as good candidates to bring about such an assembly.

3.3.1 Metal Nanocrystals

Dodecanethiol-capped Ag nanocrystals, prepared by the Brust method self-assemble into two-dimensional arrays (see Fig. 3.7) [156]. From the separation distance of 1.5 nm seen between the nanocrystal surfaces, one can deduce that the alkane chains projecting out of neighboring nanocrystals are interdigitated. The ordered arrangement of nanocrystals typically covers an area of several square microns. Thiol-capped Au nanocrystals prepared by the Brust method have been size-fractionated in order to organize into well-ordered two-dimensional arrays [581]. Two-dimensional arrays of noble metal nanocrystals

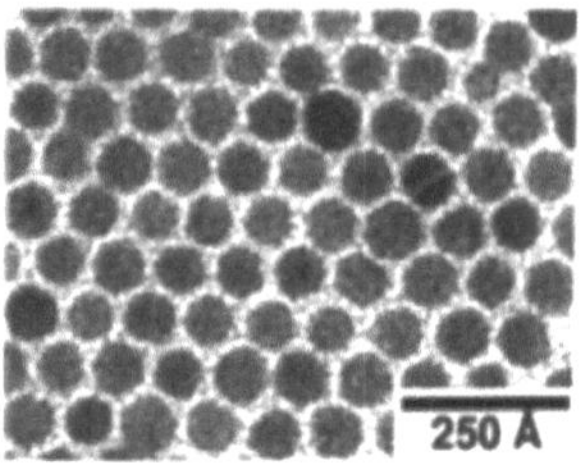

Fig. 3.7. TEM image of a two-dimensional lattice of dodecanethiol capped Ag nanocrystals. The average interparticle separation is 1.5 nm (reproduced with permission from [156])

of different sizes have been obtained starting with well-characterized hydrosols and employing a ligand exchange reaction to transfer the nanocrystals into organic medium [536, 537].

Nanocrystals that self-assemble into two-dimensional arrays often arrange themselves into superlattices consisting of a few layers of arrays. The superlattices obtained with monodisperse nanocrystals of similar sizes and capping agents but prepared using different methods have been compared [582]. Au nanocrystals obtained by the inverted micelle method self-assemble into two-dimensional arrays with face centered cubic structure, while the nanocrystals produced by the SMAD process appear to yield arrays with hexagonally close-packed layers. The difference is attributed to the change in the nature of the core from crystalline (inverted micelle) to polycrystalline or twinned in the latter case [582]. There have been attempts to organize magic nuclearity nanocrystals such as Au_{55} and Pd_{561} into two-dimensional arrays. Schmid et al. [583] obtained an ordered arrangement of Au_{55} nanocrystals on poly(ethyleneimine) functionalized TEM grids (see Fig. 3.8). The Au_{55} nanocrystals undergo a controlled growth when the capping agent is exchanged with a long-chain thiol or an amine and the resulting larger nanocrystals self-assemble into two-dimensional arrays [534, 584]. Well-ordered arrays of magic nuclearity nanocrystals, Pd_{561} and Pd_{1415}, have been successfully obtained (see Fig. 3.9) by functionalizing the surface of these nanocrystals with alkanethiols of different chain lengths [585].

Harfenist et al. [112] steered a beam of naked Ag clusters through a toluene solution containing alkane thiols and found that the clusters so capped can form extended two-dimensional arrays. Kimura and coworkers [586] obtained two-dimensional arrays of Au nanocrystals from a hydrosol. This is the first report of a two-dimensional organization occurring in an aqueous medium. An interdigitated layer of octanethiol and ethanol is thought to be responsible for bringing about such an organization. Two-dimensional arrays of alloy nanocrystals such as Au–Ag [587, 588] and Fe–Pt [297] have been obtained using thiols or amines as capping agents. Large superlattices of FeCo nanoparticles extending over several millimeters has been achieved [589].

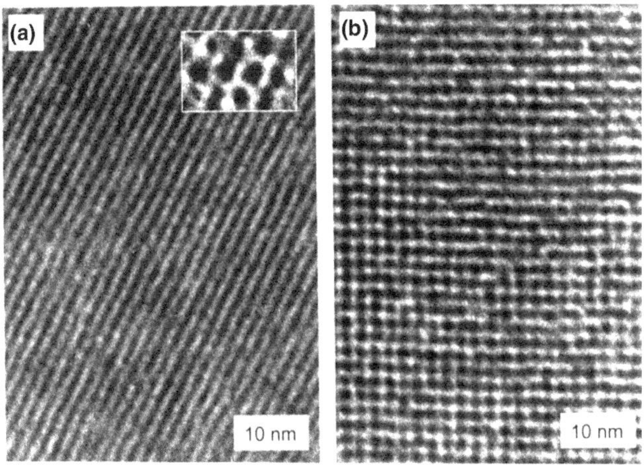

Fig. 3.8. TEM images of Au_{55} monolayers showing (**a**) hexagonal and (**b**) cubic structures. The monolayers were prepared on a polyethyleneimine functionalized carbon grid. The magnified inset in (a) shows single clusters in the hexagonal form (reproduced with permission from [583])

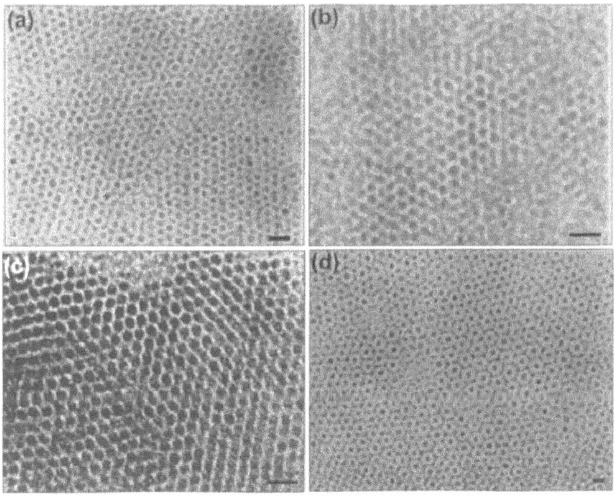

Fig. 3.9. TEM micrographs showing hexagonal arrays of thiolized Pd nanocrystals: (**a**) Pd_{561} nanocrystals capped with octanethiol, (**b**) Pd_{561} nanocrystals capped with dodecanethiol, (**c**) Pd_{1415} nanocrystals capped with octanethiol, and (**d**) Pd_{1415} nanocrystals capped with dodecanethiol (reproduced with permission from [585])

In addition to alkanethiols, long-chain fatty acids are used for ligating and assembling metal nanocrystals. A colloidal dispersion of Co nanocrystals capped with fatty acids self-assembles to yield hexagonally ordered arrays similar to those obtained with alkanethiols [162,590]. Ag nanocrystals capped with fatty acids of appropriate lengths yield cubic or hexagonal close-packed structures [591]. Two-dimensional lattices of $CoPt_3$ of various sizes have been prepared by Weller and coworkers [592] using adamentane carboxylic acid as the capping agent. An interesting aspect of this study is the use of a carboxylic acid without a long alkane chain. Ordered self-assembly of Au nanocrystals into arrays from water soluble nanocrystal micelles have been observed [593]. The method involves drying of water soluble Au nanocrystal micelles synthesized using a surfactant encapsulation technique conducted in an interfacially driven water-in-oil microemulsion process.

Wei and coworkers [594] have reported ordered two-dimensional arrays of large Au nanocrystals with diameters in the range of 15–90 nm. The large nanocrystals have been organized using resorcinarene-based capping agents. Nonspherical nanocrystals also possess the ability to self-assemble into two-dimensional arrays. Hexagonal Pt as well as elongated silver nanocrystals have been organized into ordered arrays [536, 537, 595] using alkanethiols as capping agents. Highly ordered superlattices of iron nanocubes have also been prepared [596]. An array of cube-shaped Sn nanocrystals was obtained by exploiting the affinity of an acid to its conjugate base using hexadecylamine and the hexadecylamine–HCl adduct. Superlattice formation is supposed to occur rightaway in solution rather than upon evaporation [597].

Ordered two-dimensional lattices containing thiolized spherical Au particles of two different sizes have been reported by Kiely et al. (see Fig. 3.10) [598]. The arrangement of the nanocrystals in these lattices corresponds to the predicted atomic arrangement in metal alloys with similar radius ratios. Arrays corresponding to both AB and AB_2 prototypical lattices have been obtained. An AB_5 type of lattice was observed when a solution containing $CoPt_3$ nanocrystals of diameters 4.5 and 2.6 nm was evaporated on a substrate [592]. Such an organization is also obtained starting with nanocrystals of different elements with appropriate radius ratios. Thus, two-dimensional AB and AB_2 type arrays consisting of alternate rows of thiol-capped

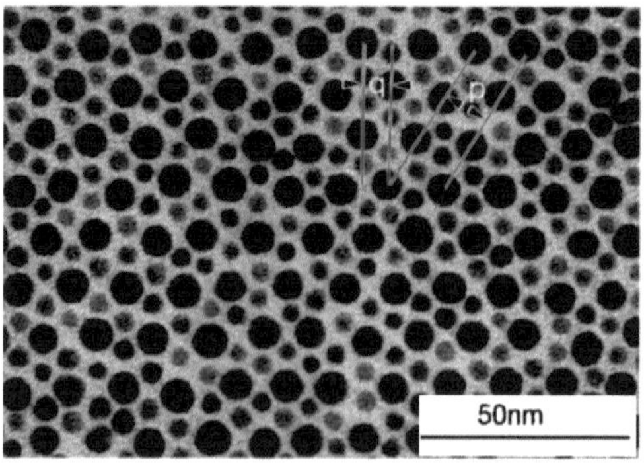

Fig. 3.10. TEM images of an array of Au nanocrystals with the AB structure. The radius ratio of the nanocrystals is 0.58 (reproduced with permission from [598])

Au and Ag nanocrystals of different sizes have been reported [599]. An AB type of superlattice has been obtained starting with oleic acid-capped Fe and dodecanethiol-capped Au nanocrystals [600].

3.3.2 Semiconductor and Oxide Nanocrystals

Size-fractionated TOPO-capped CdSe nanocrystals prepared by the method of Murray and coworkers [326] order spontaneously into two-dimensional arrays and superlattices upon evaporation of the solvent on a carbon grid [601,602]. TEM images showing the different facets of the two-dimensional layers in superlattices are shown in Fig. 3.11. Hexanethiol-capped PbS nanocrystals are found to organize into two-dimensional lattices. Oleic acid-capped PbSe nanocrystals after size-selective precipitation organize into extended arrays [334]. Motte et al. [603] obtained a hexagonally ordered two-dimensional array of Ag_2S nanocrystals employing the reverse micelle method. InP nanocrystals prepared by thermolysis have been organized into two-dimensional arrays [604]. Nanocrystals of PbS, Cu_2S, and Ag_2S spontaneously organize into superlattices upon precipitation, following synthesis under solvothermal conditions [605]. This is a rare example of solvothermally produced near-monodisperse nanocrystals assembling into a two-dimensional lattice.

The first observation of a two-dimensional nanocrystalline array was made unwittingly by Bentzon et al. [318] with Fe_3O_4 nanocrystals. These workers found that the ferrofluid obtained by the thermolysis of ironpentacarbonyl yielded well-ordered two-dimensional arrays of Fe_3O_4 nanocrystals upon drying over several weeks. Since then, easier methods have been devised to obtain arrays of Fe_3O_4 nanocrystals [606]. Iron oxide nanocrystals of different compositions, sizes, and shapes are reported to organize into two- and three-dimensional superlattices [504]. Two-dimensional arrays of amine-capped metal oxide nanocrystals such as Co_3O_4 have been obtained

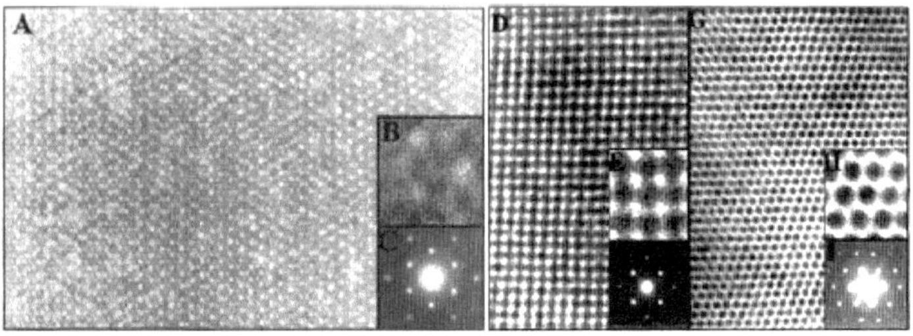

Fig. 3.11. TEM images of three-dimensional superlattices of 4.8 nm CdSe nanocrystals along (**a**) ⟨100⟩, (**b**) ⟨101⟩, and (**c**) ⟨111⟩ orientation. High-resolution images as well as selected area electron diffraction patterns are shown alongside in each case (reproduced with permission from [602])

starting from metal oxide nanocrystals prepared by the thermolysis of metal cupferrons [322].

Attempts to organize nonspherical oxide nanocrystals have also met with reasonable success. Thus, tetrahedral CoO nanocrystals are found to organize into extended two-dimensional arrays [607]. Prismatic $BaCrO_4$ nanocrystals assemble into a rectangular arrangement [513].

Arrays consisting of magnetic Fe_3O_4 and Fe–Pt nanocrystals with different sizes have been obtained. Such arrays exhibit interesting magnetic properties arising from exchange interactions between the nanocrystals. Redl et al. [608] obtained three-dimensional binary lattices consisting of magnetic γ-Fe_2O_3 and semiconducting PbSe nanocrystals in ordered two-dimensional layers. The lattices obtained possessed prototypical structures such as AB_2, AB_5, and AB_{13} (see Fig. 3.12). Nanocrystals of a variety of materials with different sizes and functionality can be made to self-assemble into ordered binary superlattices (as opals or colloidal crystals) retaining size tunable properties. Typically, monodisperse nanocrystals of PbS, PbSe, Fe_2O_3, Au, Ag as well as Pd have been used. The superlattices are isostructural with known

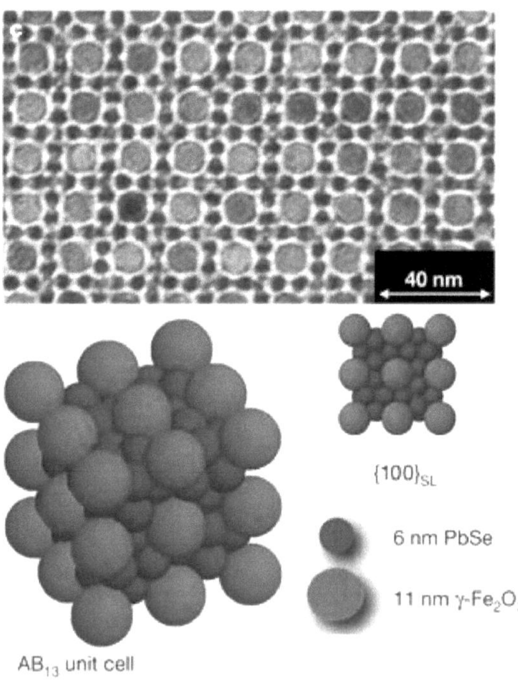

Fig. 3.12. TEM image showing a superlattice of PbSe and γ-Fe_2O_3 nanocrystals. The projection corresponds to ⟨100⟩ plane of the AB_{13} structure. The AB_{13} lattice and the structure of the ⟨100⟩ plane are illustrated (reproduced with permission from [608])

atomic lattices such as NaCl, CuAu, etc., emphasizing the parallelism between nanoparticle assemblies and atomic crystals [609].

3.3.3 Other Two-Dimensional Arrangements

Neat Au nanocrystals as well as other nanocrystals of various sizes have been organized into two-dimensional lattices using a protein (chaperonin) template [610]. Two-dimensional layers of nanocrystals with some degree of short-range order can be obtained by using the air–water interface as a template. Application of surface pressure helps to improve the quality of the monolayer as well as facilitates transfer of the film to a substrate. Such studies are typically carried out using a Langmuir–Blodgett trough. Surface pressure–volume isotherms have been obtained with metal [611,612], semiconductor [613], and oxide [614] nanocrystals coated with hydrophobic ligands. Thiol-capped Au nanocrystals are reported to organize in the form of polydisperse opals at the air–water interface [612].

3.3.4 Mechanism of Organization

The cooperative assembly of nanocrystals described hitherto was mainly driven by entropy, the factors playing a key role in deciding the nature of organization being the nanocrystal diameter (D) and the ligand chain length (ℓ). It is observed that for a given diameter of a nanocrystal, the packing changes swiftly as the length of the thiol ligand is varied. Based on a study of the organization of Pd nanocrystals of diameters in the range of 1.8–6.0 nm with alkane thiols of various chain lengths, a phase diagram in terms of D and ℓ, shown in Fig. 3.13, has been derived [585]. In Fig. 3.13, the intensity of grey shade is representative of the disorder in nanocrystalline organization. The light area corresponds to ordered arrays and the dark area to disordered arrangements. Although entropy driven, the organization is treated as being due to soft spheres rather than hard spheres.

Korgel et al. [156, 615] observed that changing the nature of the solvent used in the evaporation step results in dramatic changes in the arrays obtained. When toluene was used as the solvent, extended two-dimensional arrays made of a single layer of nanocrystal were obtained. With the addition of increasing amounts of ethanol to toluene, the nanocrystals exhibited an increasing affinity to form multilayered structures with one nanocrystal collapsing on another. To account for these observations, Korgel et al. [156,615] proposed a soft sphere model taking the interparticle interaction into consideration. Accordingly, a ligated nanocrystal allows for penetration of each other's ligand shell up to a certain extent. In this model, the total potential energy, E, between a pair of nanocrystals, is considered to be a result of two types of forces between the nanocrystals,

$$E = E_{\text{steric}} + E_{\text{vdW}} \tag{3.1}$$

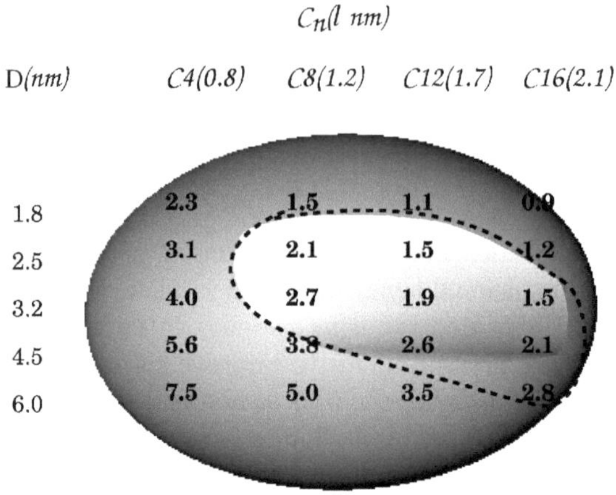

Fig. 3.13. The D–ℓ phase diagram for Pd nanocrystals caped with different alkanethiols. The mean diameter, D, was obtained from the TEM measurements on as prepared sols. The length of the thiol, ℓ, is estimated by assuming an all-*trans* conformation of the alkane chain. The thiol is indicated by the number of carbon atoms, C_n. The bright area in the middle encompasses systems that form close-packed organizations of nanocrystals. The surrounding darker area includes disordered or low-order arrangements of nanocrystals. The area enclosed by the *dashed line* is derived from calculations from the soft sphere model (reproduced with permission from [585])

van der Waals interaction due to the polarization of the metal cores constitutes the attractive term (E_{vdW}) and the steric interaction between the thiol molecules on the two surfaces forms the repulsive term (E_{steric})

$$E_{\mathrm{vdW}} = \frac{A}{12}\left\{\frac{D^2}{\tau^2 - D^2} + \frac{D^2}{\tau^2} + 2ln\left[\frac{\tau^2 - D^2}{\tau^2}\right]\right\} \tag{3.2}$$

$$E_{\mathrm{steric}} = \frac{50D\ell^2}{(\tau - D)\pi\sigma_{\mathrm{a}}^3} kTe^{-\pi(\tau-D)} \tag{3.3}$$

where, τ is the interparticle distance, A the Hamaker constant, σ_{a}, the footprint or area occupied by the thiol molecule on the nanocrystal surface. In the case of Pd nanocrystals coated with alkane thiols in toluene, it was found that the total energy is attractive over a range of interparticle distances, the magnitude increasing with fall in distance (see Fig. 3.14). The lowest points in the total energy curve refer to the maximum stabilization energy achievable for a given set of D and ℓ values. In the case of Pd nanocrystals with $D = 4.5$ nm, stabilization energies of 17 and 2 meV are obtained for particles coated with octanethiol and dodecanethiol, respectively. When the stabilization energies have moderate values, comparable to the thermal energy, ordered organizations are to be expected. If the stabilization energy is not favorable, collapsed monolayers of nanocrystals or loosely packed structures would be seen. The favorable regime is the area encircled by a dashed line in the phase diagram

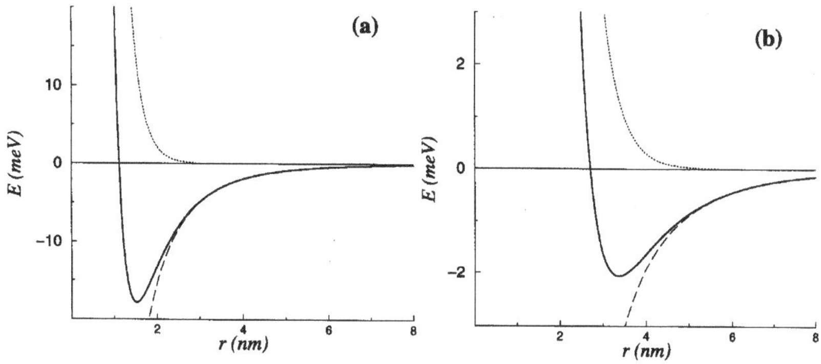

Fig. 3.14. Variation of the attractive and repulsive components and the total potential energy with the separation distance between two Pd nanocrystals of 4.5 nm diameter coated with (**a**) octanethiol and (**b**) dodecanethiol (reproduced with permission from [585])

in Fig. 3.13. A reasonable degree of correspondence is seen between the model and the experimental observations. Clearly, the interdigitation of thiol molecules plays a major role in attributing hardness to the ligated nanocrystal, which in turn decides the nature of the two-dimensional organization. This treatment is easily extendable to other metal and semiconductor nanocrystals. The observation of two-dimensional arrays with nanocrystals of different diameters clearly poses a challenge to this notion (see Fig. 3.10). The prototypical AB and AB_5 types of lattices found in these arrays are traditionally associated with the packing of hard spheres.

Bigioni et al. [616] have used video microscopy to study the process of assembly of Au nanocrystals into extended two-dimensional arrays. They suggest that the morphology of the drop-deposited nanocrystal films is controlled by evaporation kinetics and particle interactions with the liquid–air interface. In the presence of an attractive particle interface interaction, rapid early stage evaporation dynamically produces a two-dimensional solution of nanocrystals at the liquid–air interface, from which nanocrystal islands nucleate and grow upon further evaporation.

3.4 Three-Dimensional Superlattices

Multilayer assemblies of nanocrystals of CdSe are generally fragile and are not suited for use in functional devices. Assemblies involving alternate layers of nanocrystals and linkers are relatively robust. The method of layer-by-layer deposition has received a good deal of attention over the last few years, since it provides a convenient low-cost means to prepare ultrathin films of controlled thickness, suited for device applications. In a typical experiment, one end of a monolayer forming a bifunctional spacer, is tethered to a flat substrate such

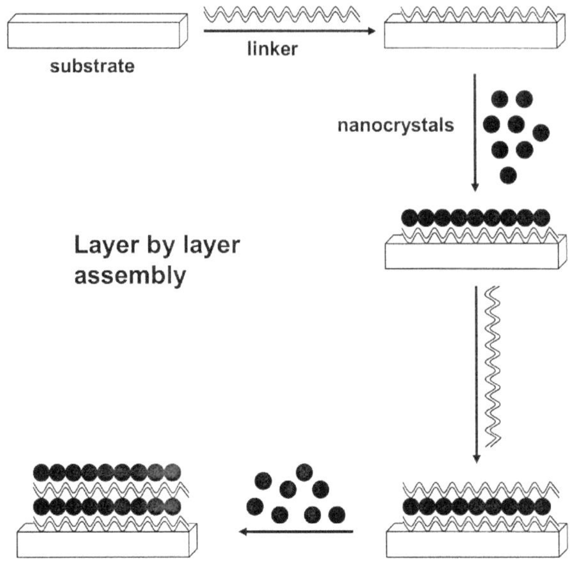

Fig. 3.15. Schematic illustration of the layer-by-layer deposition process

as gold, aluminium, indium tin oxide, or glass, leaving the other end free to anchor nanocrystals. Subsequent layers can be introduced by dipping the substrate sequentially into the respective spacer molecule solution and the nanocrystal dispersion, with intermediate steps of washing and drying (see schematic illustration in Fig. 3.15). The formation of a multilayer assembly is monitored by spectroscopic and microscopic methods.

Alkanedithiols have been used as linkers to build multilayer assemblies on Au substrates. Brust et al. [617] reported the formation of multilayers of Au nanoparticles using dithiol crosslinkers. Layer-by-layer deposition of particle arrays has been examined by UV–visible spectroscopy and ellipsometry. Multilayers of CdS nanocrystals prepared by the reverse micelle technique have been obtained on Au surfaces by the use of dithiol linkers [618]. Layer-by-layer deposition has also been carried out to prepare multilayer assemblies with alternating layers of metal nanocrystals of different sizes such as Au–Pt–Au or metal and semiconductor layers such as Au–CdS–Au [619]. The formation of multilayers was established by means of X-ray photoelectron spectroscopy, X-ray diffraction and scanning tunneling microscopy (see Fig. 3.16). There are many advantages in the use of dithiols or similar crosslinkers that bind covalently to a metal substrate and at the other end to nanocrystal. The layers consist of regularly arranged particles with high surface coverage and the interspersing layers have a well-defined structure and thickness. Disadvantages include long deposition time per layer ($\sim$12 h), opacity of the substrate, and the presence of defects such as pin holes in the dithiol layer that thwart charge transport studies. The use of dithiol linkers also restricts one to use organosols.

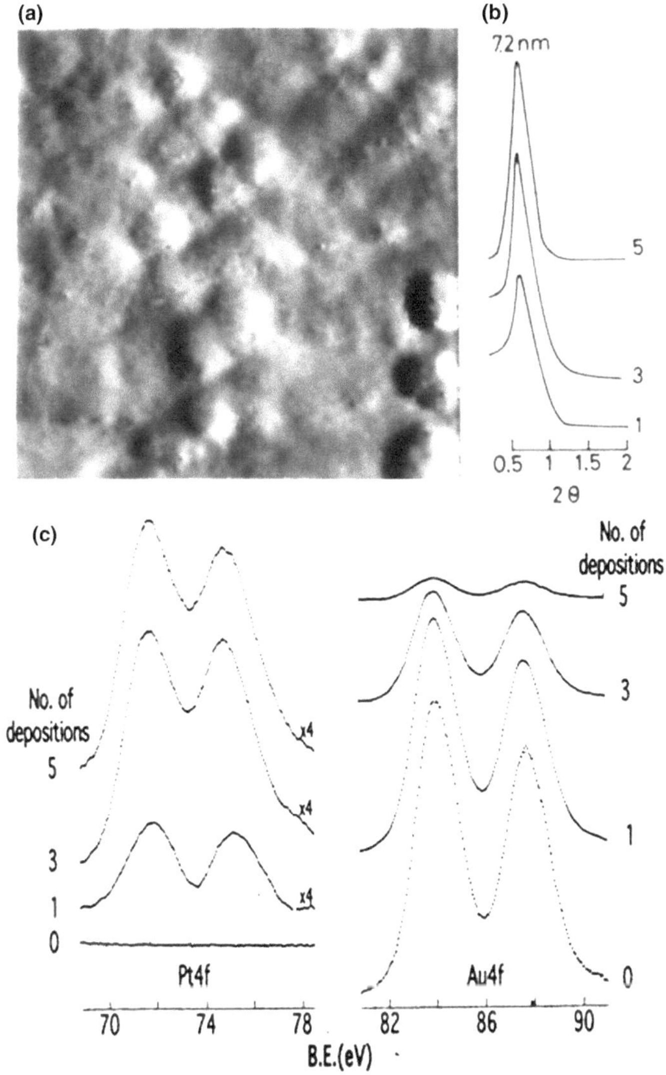

Fig. 3.16. Multilayer deposition of Pt (5 nm) nanocrystals on a polycrystalline Au substrate. After each deposition, the structure was characterized by STM, X-ray diffraction as well as by XPS. (**a**) STM image obtained after the second deposition showing the presence of regular arrays of nanoparticles with an interparticle spacing of 2 nm, extending over 300 nm, corresponding to the size of a typical flat terrace on the substrate (**b**) X-ray diffraction pattern of the arrays after the first, third, and fifth depositions exhibiting a low-angle reflections with the d-spacings reflecting the particle diameter and the interparticle distance. (**c**) X-ray photoelectron spectra in the Pt(4f) and Au(4f) regions for the 5 nm Pt/Au system. The intensity of the Pt(4f) feature increases with the number of depositions, accompanied by a decreases in the Au(4f) intensity as the substrate gets increasingly shadowed due to the limited escape depth of the photoelectrons (adapted with permission from [619])

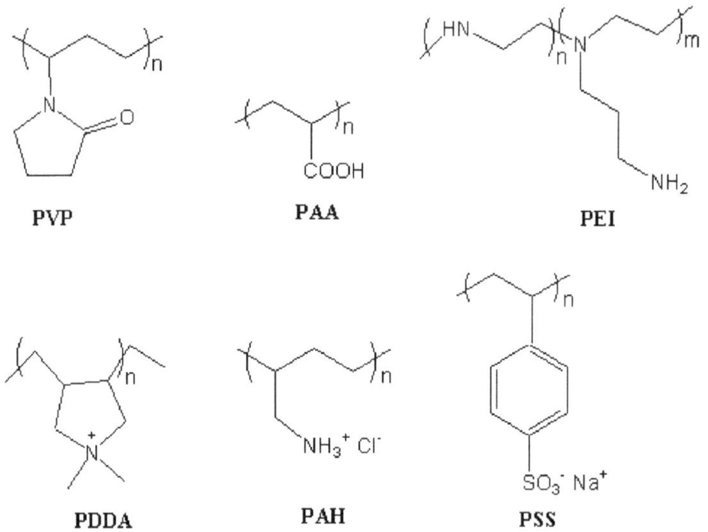

Fig. 3.17. The structures of commonly used polyelectrolytes. PEI, poly (ethyleneimine); PDDA, poly(diallyldimethylammonium chloride); PSS, poly (styrenesulfonate); PAH, poly(allylamine hydrochloride); PAA, poly(acrylic acid); and PVP, poly(vinylpyrrolidone)

Polyelectrolytes such as poly(diallyldimethylammonium chloride) (PDDA) and polyethyleneimine (PEI) are used as alternative interspacers to build up multilayers of nanocrystals. The nanocrystals bind to the polyelectrolyte layer due to electrostatic interactions. A higher degree of disorder exists in each nanocrystalline layer. However, adsorption of nanocrystals can be carried out in only a few minutes. The structures of the commonly used polyelectrolytes are shown in Fig. 3.17. By the use of alternating layers of poly(styrene sulfonate)sodium salt (PSS) and poly(allylamine hydrochloride) (PAH), multilayers consisting of Au nanocrytals were obtained on PEI-coated Si or float glass substrates. A layer of cationic PSS is deposited on the PEI covered substrate. Anionic citrate-capped Au nanoparticles adhere to the cationic PAH layer by electrostatic interactions. Another layer of anionic PSS renders the substrate suitable for further deposition of PSS and Au layers. The deposition was followed in this study by means of X-ray reflectivity and UV–visible spectroscopy measurements [620]. PAH and other polycationic polymers can be directly adsorbed on mica. On Si or glass substrates, PEI or aminosiloxanes are used as base layers to initiate the multilayer deposition process [621]. Fendler and coworkers [622] has used polycations to bring about the organization of semiconducting CdS, TiO_2, and PbS nanocrystals on quartz, Au and Teflon substrates. Si nanoparticles of 12 nm diameter self-assemble on adsorption of various gases of amines, hydrocarbons, and aldehydes [623]. PbTe nanocrystals have been assembled to obtain glassy solids and superlattices by varying the deposition conditions. The short- and long-range structural order was analyzed using HRTEM and Small Angle X-ray Scattering (GISAXS) [624].

Metal nanocrystals capped with an oxide layer such as silica or titania are useful ingredients of multilayered structures. The surface charge of the particles depends on the isoelectric point of the oxide shell and can be varied by choice of pH, etc. In aqueous solutions, silica is negatively charged. Layer-by-layer assembly of Au nanocrystals coated with silica has been obtained with PDDA as the crosslinker. On the other hand, titania-coated nanocrystals can be organized using a polyanion such as PAA. Besides polyelectrolytes, small molecules with multiple charges can be used as crosslinkers. For example, Willner and coworkers [625–628] have carried out extensive studies on multilayers of Au and Ag nanocrystals prepared by crosslinking electroactive cyclophanes based on the 4,4'-bipyridinium cation.

In addition to electrostatic interactions, weak interactions such as hydrogen bonding and acid–base interactions are used to bring about multilayered organization. Au nanocrystals capped with 4-mercaptobenzoic acid have been deposited using PVP as the crosslinker [629]. Murray and coworkers [630] have made use of acid–base interactions to build multilayer assemblies of Au nanocrystals capped with a mixed layer of hexanethiol and mercaptoundecanoic acid or 4-aminothiophenol. The carboxylic acid capped particles were organized by the use of PAH interlayers while PSS was used to organize the amine-capped nanocrystals. UV–visible spectra showing layer-by-layer assembly of Au nanocrystals is shown in Fig. 3.18. Ag nanocrystals coated with graphite oxide layer have been shown to adsorb to polycation layer due to

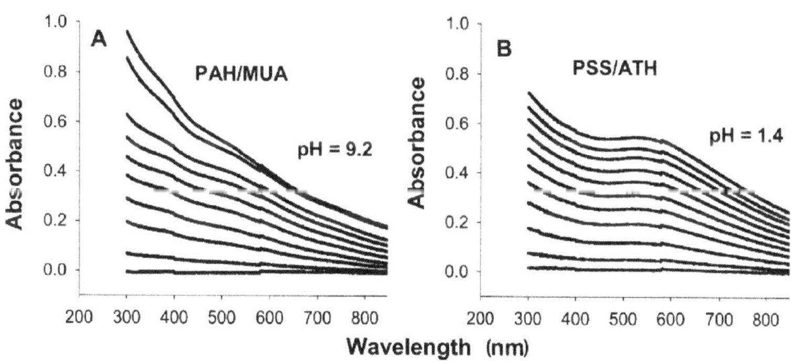

Fig. 3.18. UV–vis spectra showing layer-by-layer growth of polymer/Au nanocrystals on thiol-functionalized glass slides. (**a**) PAH/mercaptoundecanoicacid multilayer film formed by alternately exposing the slide to a solution of poly(allylamine hydrochloride) and another solution of a Au nanocrystals with a mixed ligand shell consisting of hexanethiol and mercaptoundecanoic acid. (**b**) PSS/ATH multilayer film formed by alternately exposing the slide to a solution of poly(styrene sulfonic acid) and another solution of Au nanocrystals with a mixed ligand shell consisting of hexanethiol and 4-mercaptophenylamine (ATH) (reproduced with permission from [630])

electrostatic interaction between the carboxyl or the hydroxyl groups and the cationic backbone of the polymer [549]. Poly(diallyldimethylammonium chloride)(PDDA), CdSe spaced with PPV have been prepared.

3.5 Superclusters

Theoretical considerations suggest that self-similarity in metal nanocrystal organization should manifest itself in the form of giant clusters whose shape and size are direct consequences of the nanocrystals themselves [631]. The invariance of the shell effects in metal nanocrystals with scaling is shown schematically in Fig. 3.19. Thus, Pd_{561} nanocrystals would be expected to self-aggregate into a giant cluster of the type $(Pd_{561})_{561}$ under suitable conditions. The monodisperse nature of the magic nuclearity nanocrystals is thought to be important in assisting the self-aggregation process. Formation of such clusters has been observed in the mass spectra of magic nuclearity Au_{55} nanocrystals. Secondary ion mass spectrometry has revealed the presence of species with large m/z values and these were attributed to $(Au_{13})_{55}$ giant clusters [632]. The giant clusters so obtained have, however, not been isolated or imaged. One such observation was made in the case of Pd_{561} nanocrystals where the PVP

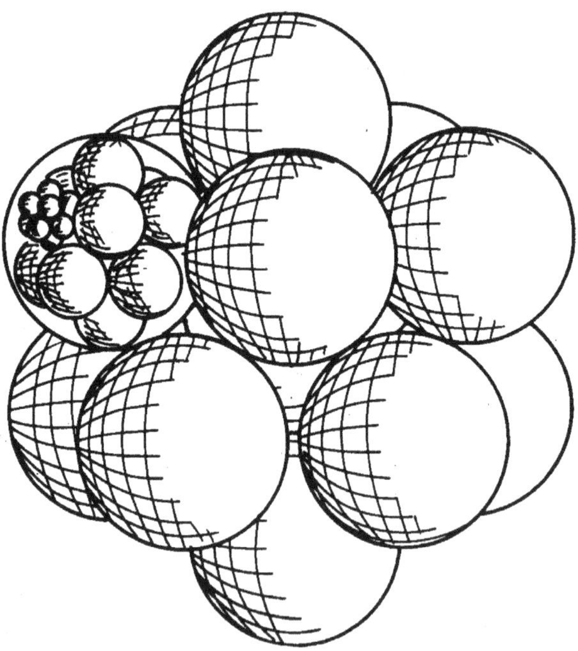

Fig. 3.19. *Self-similarity.* Schematic illustration of the formation of a cluster of metal nanocrystals (supercluster) and a cluster of superclusters. The size effects operating in nanocrystals could be invariant to scaling

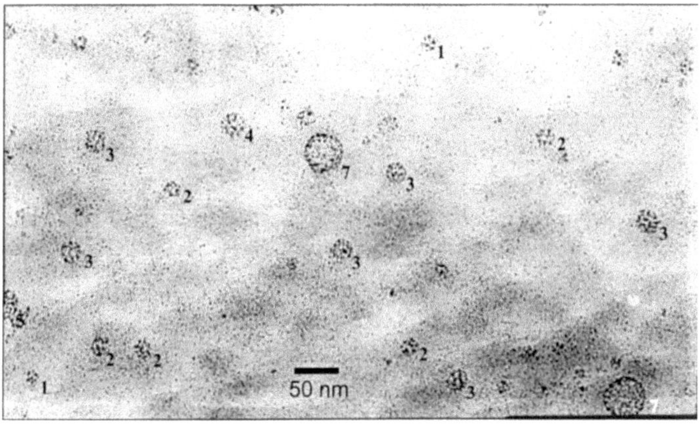

Fig. 3.20. TEM micrograph showing the giant clusters comprising Pd_{561} nanocrystals. The specimen was prepared by the slow evaporation of a PVP–Pd_{561} hydrosol. The numbers beside the aggregates indicate the estimated number of closed shells in each aggregate (reproduced with permission from [633])

covered nanocrystals aggregated to form giant clusters [633]. The TEM image in Fig. 3.20 is revealing. Pd nanocrystals in the form of dense aggregates are seen in several areas of the grid. Estimates of the number of nanocrystals that make up such aggregates yield numbers corresponding to magic nuclearities. It is possible that the formation of the giant clusters is facilitated by the polymer shell that encases them. Unlike in the case of Pd_{561} nanocrystals coated with alkanethiols, which self-assemble to form ordered arrays, the polymer shell may effectively magnify the facets of the metallic core, thereby aiding a giant assembly of the nanocrystals.

3.6 Colloidal Crystals

It has not been possible hitherto to prepare single crystal made up of nanocrystals. There is, however, a tendency of monodisperse nanocrystals to arrange into ordered three-dimensional arrays extending to a few microns. By tuning the crystallization conditions, crystallites of micrometer dimensions consisting of thousands of Au_{55} nanocrystals have been prepared [634]. Microcrystallites of CdSe [601] and Au [582] have been similarly obtained. Microcrystals of mercaptosuccinic acid capped Au nanocrystals have been obtained from an aqueous medium [635], through an assembly mediated by hydrogen bonding interactions between the ligand molecules. Weller and coworkers [636] have devised a highly successful method of producing these microcrystals using what is called a three-layer technique of controlled oversaturation (see Fig. 3.21). In this method, a nonsolvent such as methanol is allowed to diffuse slowly through a buffer solvent layer such as propanol to a toluene layer containing

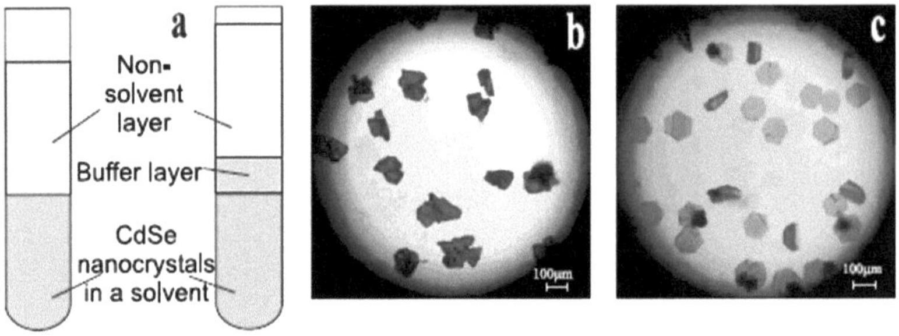

Fig. 3.21. (**a**) Schematic illustration of the three-layer technique of controlled oversaturation. (**b**) Photograph of CdSe microcrystals prepared by two-layer technique (left test tube). (**c**) Photograph of CdSe microcrystals prepared by three-layer technique (right test tube). The crystallites in (c) have better defined facets (reproduced with permission from [637])

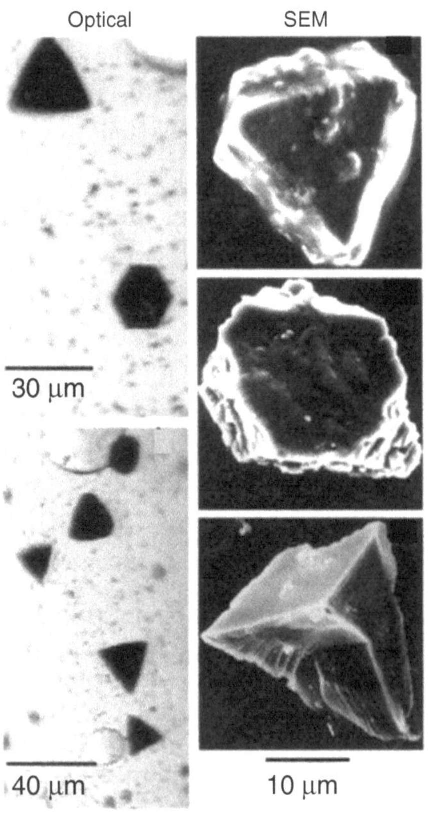

Fig. 3.22. Photographs and SEM images of microcrystals of Fe–Pt nanocrystals (reproduced with permission from [636])

nanocrystals. By carrying out this process in long narrow tubes, microcrystals of nanocrystals are obtained after a few weeks. This technique has been used to prepare microcrystals of $CoPt_3$, Fe–Pt, and CdSe nanocrystals (see Fig. 3.22) [574, 636, 637]. Careful experiments reveal that the arrangement of nanocrystals in such crystallites are polymorphic. In the absence of single crystals, it is believed that such crystallites could prove to be the best candidates to study properties unique to such ensembles of nanocrystals.

4

Properties of Nanocrystals

The basis for size-dependent changes in the properties of nanocrystals was outlined earlier in Chap. 1. In this chapter, we shall examine this aspect in some detail with examples. It is useful to categorize sizes of nanocrystals into different regimes specific to the different properties, beyond which size-dependence would not be relevant. The schematic in Fig. 4.1 illustrates this aspect. All the regimes begin essentially with small clusters (≥1 nm), but the upper limits are different. There are phenomenological approaches to the problem of size dependence [638]. Jortner suggests that a cluster property, G, can be represented by an universal scaling law involving N or R. Thus,

$$G(R) = G(\infty) + aR^{-\alpha_1}, \tag{4.1}$$

$$G(N) = G(\infty) + aN^{-\beta_1}. \tag{4.2}$$

Here, N and R represent nuclearity and radius, respectively. The values of the exponents α_1 and β_1 are normally 1 and $\frac{1}{3}$, respectively. The scaling law describes the experimental observations on ionization energy and charging energy reasonably well. When the size of clusters becomes extremely small (say less than 10–12 metal atoms), the clusters act like molecules with different energy level manifolds.

4.1 Melting Point and Heat Capacity

"Does the melting temperature of a small particle depend on its size?" asked Lord Kelvin as early as in 1871 [639]. The consensus in those days seemed to be that it would not be the case. An initial attempt was made to examine this issue by Pawlow in 1909 [640]. The first demonstration that the melting point was indeed different in small particles is due to Takagi [641], who established by means of electron microscopy that nanoscale particles of Pb, Sn, and Bi with sizes in the range of a few nanometers exhibited lower melting temperatures. Buffat et al. [642] carried out extensive studies on the melting points

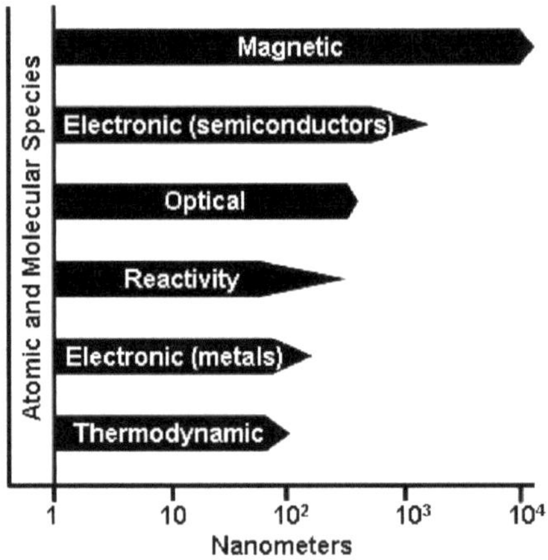

Fig. 4.1. Schematic illustration of the size-dependence of various nanocrystal properties. The property specific regimes are indicated

of Au nanocrystals by means of scanning electron diffraction technique and compared their results with previous findings. The change in the melting point can be quite dramatic, with lowering of as much as 600 K in Au nanocrystals. It is supposed that the surface atoms are more susceptible to thermal displacement and initiate the melting process due to the lower coordination. Such a surface melting process is thought to be the major cause for the lowering of melting points. Shi and others have developed theoretical models predicting a 1/r dependence for the melting behavior of nanocrystals [643, 644]. In the recent past, an entirely different melting behavior has been observed for much smaller particles of nuclearities less than 500.

Using calorimetry, the groups of Jarrold and Haberland have carried out studies on Ga and Na clusters [645–649]. The Ga clusters were found to have melting points far higher than the bulk (302.9 K). For example, the Ga_{40}^+ ions showed evidence of melting at 550 K while no melting was observed in the case of Ga_{17}^+ even up to 720 K. In the size range of 30–55 atoms, the addition or removal of a single atom resulted in dramatic changes in the melting behavior of these clusters. The unusual stability of Ga clusters is attributed to the presence of covalent bonds in the tiny clusters [650]. The covalent bonds are replaced by metallic bonds in the bulk. Na clusters with nuclearities in the range of 40–350 atoms have been studied. Clusters with nuclearities <90, do not show a clear signature of melting, but the clusters that melt, do so at temperatures below the bulk melting temperature. The observed melting behavior of Na clusters is explained in terms of the stability of the geometric shells [651].

In addition to size-dependent melting points, size-dependent specific heats have been observed. At high temperatures, there is an increase in the specific heat of nanocrystals. For example, an enhancement of 29–53% is seen in the case of Pd nanocrystals of ~6 nm diameter in the 150–300 K range [652]. Similar but less dramatic enhancements have been seen in other metallic and semiconducting elements. There is no clear explanation for these observations. At temperatures near absolute zero, when the specific heat can be decomposed into lattice and electronic contributions, interesting effects are observed. The early work of Frölich [653] and Kubo [49] showed that the electronic contribution to specific heat is reduced up to two-thirds. Bai et al. [654] have measured the changes in specific heat of thermally evaporated Fe particles of 40 nm diameter at low temperatures. An increase in specific heat was observed, but such a change was considered to be consistent with an increase in the lattice contribution and a decrease in the electronic contribution as predicted by Frölich. Such a decrease is observed in heat capacity measurements at very low temperatures on magic nuclearity Pd nanocrystals [655].

4.2 Electronic Properties

Electronic properties of nanocrystals critically depend on size. This aspect is aptly put forth in the quest "How many atoms make a metal?". It is clear that as the size of metal nanocrystals is reduced, the accompanying changes in the electronic structure render them insulating. This transition, called the size-induced metal–insulator transition (SIMIT), has evoked much interest from chemists and physicists alike. A SIMIT is manifested in experiments that measure the electronic band structure and atomistic properties such as ionization energy.

In small metal particles containing up to a few hundred atoms, the electronic properties are entwined with changes in the structure and bonding. As outlined in Chap. 1, issues arising from the closure of electronic and geometric shells also play a part in determining the electronic structure of small metal particles. A large part of the understanding of bare metal clusters follows from the measurement of ionization energy [52, 54]. Ionization energy is generally measured by means of photoionization. A cluster beam is ionized by photons from a monochromatized UV source or a tunable laser and the mass distribution of the produced ions are analyzed after considering factors such as the temperature of the clusters and their cross-sections. Such measurements have been carried out on several s, p, and transition metal clusters [52, 54]. Typical curves of ionization energy vs. nuclearity for Na and K clusters are shown in Fig. 4.2. Large jumps are seen in the ionization energy for nuclearities corresponding to the closure of electronic shells. Similar trends have been observed in other alkali, alkaline earth, and p-block metals. A noteworthy feature of the data in Fig. 4.2 is the presence of an odd–even effect in very small clusters which is in accord with the predictions by Frölich. The ionization energy of a cluster with odd nuclearity is clearly lower than its even neighbors. Another interesting aspect is revealed when one compares Figs. 4.2 and

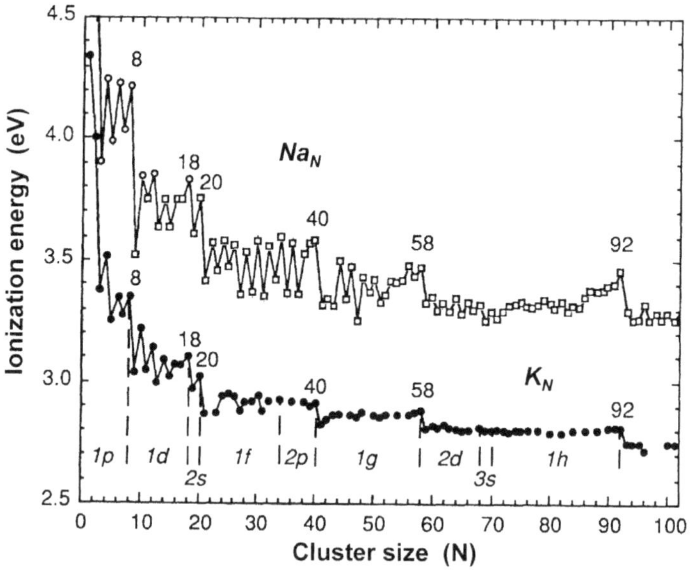

Fig. 4.2. Ionization energy of size-selected Na and K clusters as a function of nuclearity, N. The nuclearities corresponding to the closure of electronic shells are indicated (reproduced with permission from [52])

1.12. At small nuclearities, the ionization energy appears to be dominated by electronic shell effects while at higher nuclearities ($N > 2{,}000$), the structure is dictated by the geometric shell effects. The transition from the electronic to the geometric shell structure is not well understood at this point of time. Clemenger [656] opines that the dictates of geometry take over when perturbations caused by them become greater than those from electronic shell effects. He estimates that such a change should occur for $N = 1{,}640$, in agreement with experiment. Ionization energies and electronic affinities have been modeled based on electrostatic considerations (the spherical drop model) as well as semiclassical and quantum mechanical theories [52, 54].

Chesnovsky et al. [657] and Rademann et al. [658] studied mass-selected Hg clusters by photoelectron spectroscopy in the UV range. They found that a small Hg particle with atoms in the $6s^2 6p^0$ configuration is held together by relatively weak van der Waals forces and is essentially nonmetallic. As the particles grow in size, the atomic 6s and 6p levels broaden into bands and an insulator–metal transition appears to occur (see Fig. 4.3). The SIMIT occurs when the number of atoms in the cluster is around 400. Although such measurements have been attempted on binary semiconductors, only limited information could be obtained due to the strong dependence of the spectra on the geometric structure as well as effects arising from the dangling bonds on the surface of a semiconducting nanocrystal [659, 660].

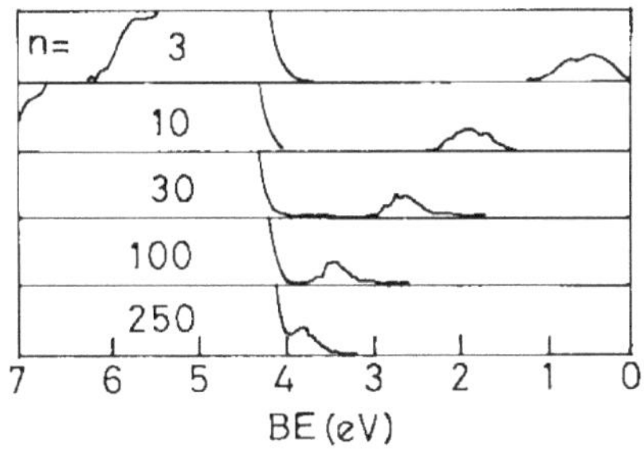

Fig. 4.3. Photoelectron spectra of Hg clusters of varying nuclearity. The 6p feature moves gradually toward the Fermi level, emphasizing that the band gap shrinks with increase in cluster size (reproduced with permission from [657])

X-ray photoelectron spectroscopy of nanocrystals has yielded useful information on the electron states, wherein changes in the electronic structure are manifested as variations in the core-level binding energies. The shifts in the binding energy of metal nanocrystals or clusters have been measured relative to the bulk for metals such as Au, Ag, Pd, Ni, and Cu by varying the particle size [661, 662]. Such experiments reveal that as the particle size decreases, the binding energy increases sharply by as much as 1.2 eV as in the case of Pd (see Fig. 4.4). The variation is negligible at large particle sizes since the binding energy would be close to that of the bulk metal. The increase in the core-level binding energy in small particles is due to the poor screening of the core holes as seen by the outgoing photoelectron. This is taken as a manifestation of SIMIT. Au^{197}Mössbauer studies of Au and Pt nanocrystals as well as cluster compounds have revealed metal-like isomer shifts for the surface atoms in larger particles. The surface atoms in smaller clusters, on the other hand, exhibited isomer shifts characteristic of partially oxidized metal ions [663]. In the case of thioglycerol-capped CdS nanocrystals, X-ray photoelectron spectroscopy has yielded an estimate of the diameter of the nanocrystals [664]. The spectra contained signatures of three S species, unlike bulk CdS which contains only one kind of S species. These have been identified as originating from the core, the surface layer, and the capping agent. From the ratio of intensities of the former two species, the size of the nanocrystals was deduced.

The occupied and the unoccupied energy levels around the Fermi level can be probed employing UPS and Bremsstrahlung Isochromat (BI) spectroscopic techniques. A study of Pd nanocrystals (Fig. 4.5) has shown that for large particles (~5 nm), the UP spectrum is similar to that of the bulk metal, with considerable intensity (density) of the Pd(4d) states at E_{f}. These nanocrystals therefore are clearly metallic [661]. For small particles (<2 nm), the intensity at E_{f} decreases markedly and is accompanied by a shift in the intensity maximum to higher binding energies and narrowing of the 4d-related spectral features. The BI spectra of small nanoparticles show that the empty

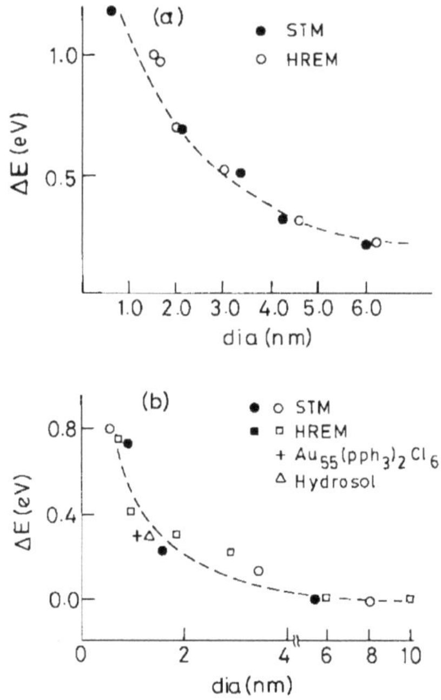

Fig. 4.4. Variation of the shifts in the core-level binding energies (relative to the bulk metal value) of (**a**) Pd and (**b**) Au nanocrystals, with the average diameter. The diameters were obtained from HREM and STM images. In the case of Au clusters, data for a colloidal particle and an Au_{55} compound are also shown (reproduced with permission from [661])

4d states have little intensity at E_f. With the increasing particle size, new states emerge closer to E_f making the spectrum closer to that of the bulk metal. Similarly, the BI spectra of Ag nanoparticles showed a distinct feature which moves toward E_f with increasing size of the particles, accompanied by an increase in the intensity of the 5s band. Thus, both UP and BI spectra reveal that the density of states around E_f is depleted in the case of small particles and that the large nanocrystals effectively mimic the metallic state of the bulk. A SIMIT is indeed implied by these measurements.

Theoretical investigations of the electronic structures of metal nanoparticles have thrown light on size-induced changes. Rosenblit and Jortner [665] calculated the electronic structure of a model metal cluster and predicted that electron localization should occur in a cluster of ~0.6 nm diameter. Clusters with magic nuclearity have been studied extensively assuming cuboctahedral, octahedral, and icosahedral structures. A molecular orbital calculation on the Au_{13} cluster [666] shows that the icosahedral structure undergoes Jahn–Teller distortion while the cuboctahedral structure does not distort. The onset of the metallic state is barely discernible in the Au_{13} cluster. Relativistic density functional calculations of gold clusters, with n = 6-147 show that the average

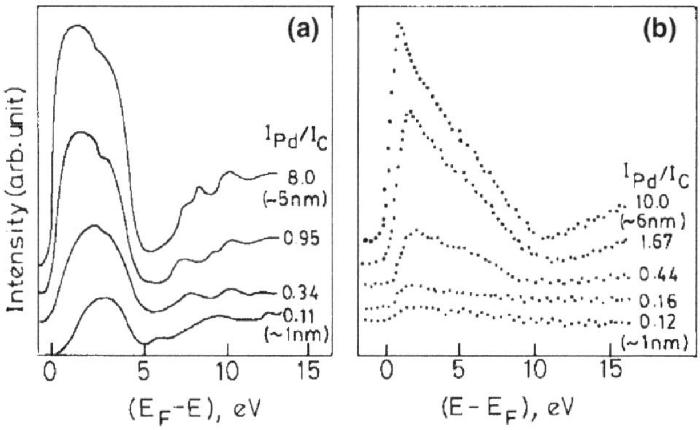

Fig. 4.5. Electronic structure of small Pd clusters near the Fermi level. (**a**) He II UP spectra show a decrease in the intensity of the occupied 4d band with the decreasing cluster size. This is accompanied by a shift in the intensity maximum to higher binding energies and a narrowing of the spectral features. (**b**) BI spectra show similar effects in the empty 4d states above the Fermi level. $I_{\mathrm{Pd}}/I_{\mathrm{C}}$ values (ratios of XPS core-level intensities of Pd and the graphite substrates) are used as a measure of the coverage or the cluster size. The approximate cluster diameter is indicated in extreme cases (reproduced with permission from [661])

interatomic distance increases with the nuclearity of the cluster [667]. The HOMO–LUMO gap decreases with particle size from 1.8 eV in Au_6 (0.5 nm diameter) to 0.3 eV in Au_{147} (2 nm diameter) [22]. The HOMO–LUMO gap increases from 0.5 eV for Al_2 to 2 eV for Al_{13}; the gap is around 0.25 eV in Au_{55}, decreasing to 0.1 eV for Au_{147}.

STS provides a direct method to determine both the occupied and the unoccupied energy levels of a nanocrystal. The energy gaps of metal nanocrystals of different diameters have been measured by means of scanning tunneling microscopy carried out in vacuum [668]. It was found that, below a cluster diameter of 1 nm, an energy gap occurs, the value of which increases with the decrease in cluster size. This value, a measure of the band gap, reaches a value up to 70 meV at small cluster sizes (see Fig. 4.6). Employing a slightly different method of tunneling spectroscopy, Sattler and coworkers [669] find that Si nanocrystals larger that 1.5 nm exhibit metal-like behavior with no bandgap. This is attributed to the presence of dangling bonds on the surface, which populate the mid-gap region, virtually eliminating the gap. STS measurements have been carried out on semiconductor nanocrystals such as CdS and InAs at low temperatures [670–674]. An increase in the tunneling gap is generally seen with a decrease in size (see Fig. 4.7). In addition to measuring changes in the gap, tunneling spectra throw light on the nature of bands. For example, the doublets close to zero in the positive bias region and the sextets in the negative bias region in Fig. 4.7 reveal the s and p character of the HOMO and LUMO levels.

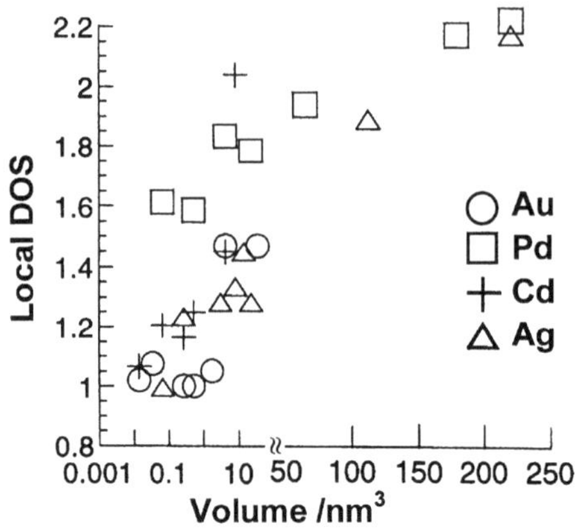

Fig. 4.6. Variation of the nonmetallic band gap with nanocrystal size in metal nanocrystals. The bandgaps were obtained based on scanning tunneling spectroscopy measurements (reproduced with permission from [668])

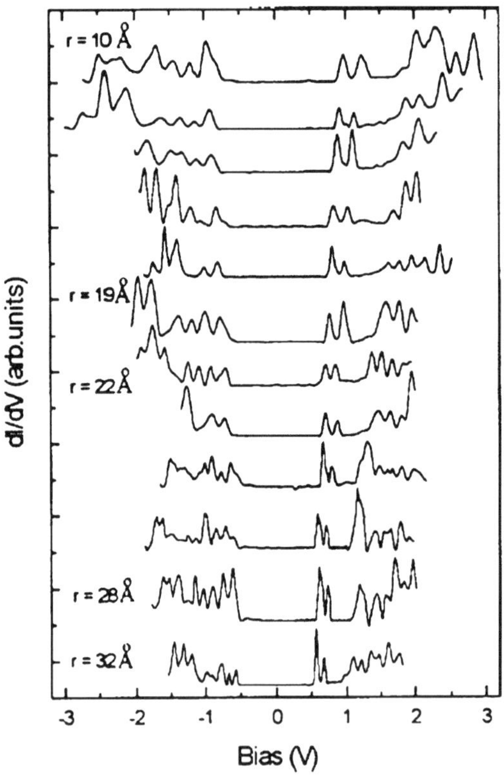

Fig. 4.7. Size evolution of the tunneling dI/dV vs. V characteristics of InAs nanocrystals acquired at 4.2 K. The spectra are displaced vertically for clarity. Representative nanocrystal radii are denoted (reproduced with permission from [674])

Tunneling spectroscopic measurements have also opened the door for the realization of single electron circuits. It is possible to design circuits where the characteristic charging energies of a cluster is manifest as a Coulomb blockade or staircase. The possible effects of single electron charging on the I–V characteristics were realized as early as the early 1950s [62, 65, 675]. It is due to the pioneering experiments of Ammen and coworkers [676] on Au nanocrystals in 1991 that a clear Coulomb staircase was experimentally established. In order to observe a blockade or a staircase behavior, the nanocrystals need to be isolated from the electrodes by insulating layers. Thus, the circuits adopt an MINIM (Metal–Insulator–Nonmetal–Insulator–Metal) configuration. The characteristics of an MINIM junction can be modeled on the basis of an equivalent circuit, consisting of two resistors R_1 and R_2 and two capacitors C_1 and C_2 in parallel. The total capacitance C_{T} and the resistance R_{T} are given by

$$R_{\mathrm{T}} = R_1 + R_2 \tag{4.3}$$

$$1/C_{\mathrm{T}} = 1/C_1 + 1/C_2 \tag{4.4}$$

At high values of resistance (hundreds of MΩ) and low values of capacitance ($<10^{-18}$ F), a voltage of $e/2C_2$ is required before the current begins to flow through the circuit [64]. This region of current exclusion is specifically called Coulomb blockade. The breaking of the tunneling barrier increases the occupancy of electrons at the junction of two capacitors to 1. A further increase in occupancy requires an increase of e/C_2 in the charging potential. If the electron resides long enough at the junction, it would provide a feedback, preventing another electron from reaching the junction till a charging potential of $3e/2C_2$ is reached. This would lead to the observation of a step-like behavior (Coulomb staircase) in the I–V spectra. The residence time of the electron is proportional to the ratio of the resistances and capacitances. Thus, steps are visible when $C_1 \ll C_2$ and $R_1 \ll R_2$. The earlier treatment assumes that the charging energy associated with this circuit ($e/2C$) is greater than kT. While Coulomb blockade may be observed when the earlier conditions are satisfied, the design of a circuit to exhibit Coulomb staircase behavior is somewhat tricky considering the strong asymmetry required in the values of the resistances and capacitances.

Tunneling spectroscopic methods provide a ready means of varying the characteristics of the second junction in an MINIM configuration by varying the tunneling gap and have therefore become the methods of choice [677]. For example, Coulomb staircase behavior is observed at room temperature in the tunneling spectra of Pd and Au nanocrystals in the size range of 1.5–6.5 nm (see Fig. 4.8) [678]. The charging energies obtained from the I–V spectra follow a scaling law [679] of the form, $U = A + B/d$, where A and B are constants, characteristic of the metal (see Fig. 4.9).

There have been attempts to design circuits that involve both self-assembly and exhibit single electron transport at room temperature [62–64]. It is possible to observe Coulomb blockade from Au nanocrystals linked to thiol molecules tethered to a Au surface [677]. Andres and coworkers [680] investigated

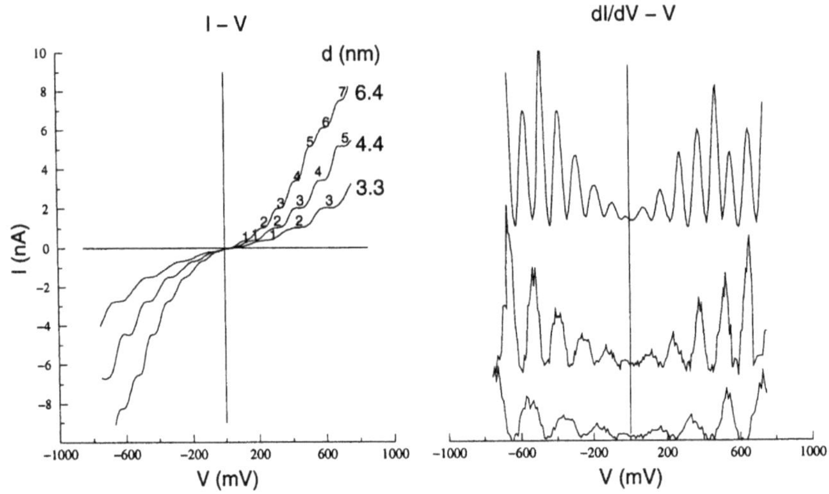

Fig. 4.8. I–V spectra of isolated PVP capped Pd nanocrystals of different diameters exhibiting Coulomb staircase behavior. The derivative spectra ($\mathrm{d}I/\mathrm{d}V$ vs. V) are shown alongside (reproduced with permission from [678])

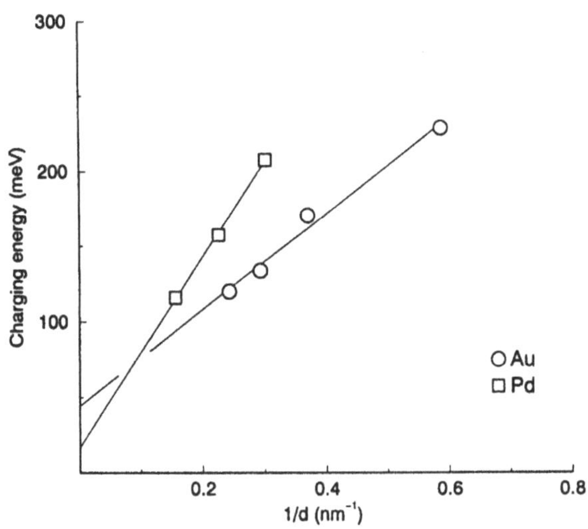

Fig. 4.9. Variation of the charging energies of Pd and Au nanocrystals with inverse diameter (reproduced with permission from [678])

the I–V characteristics of Au nanocrystals assembled on a monolayer of p-xylene-α,α'-dithiol molecules. Coulomb blockade was observed at room temperature from such a circuit. Mallouk and coworkers [681] have fabricated MINIM devices using layers of polyelectrolytes, exfoliated sheets of insulating solids, and Au nanocrystals. The device thus fabricated exhibited a Coulomb blockade behavior, the gap being tunable over a range by altering the thickness of the constituent layers. Murray and coworkers [682–685] observed Coulomb staircase behavior from an ensemble of monodisperse Au nanocrystals by electrochemical means. The steps in the I–V curves were comparable to those in individual nanocrystals obtained from tunneling spectroscopy. Electrochemical method is relatively easier for the observation of single electron transport.

Charge-transport characteristics of nanocrystalline assemblies have been addressed in a number of studies. Pellets of monodisperse nanocrystals, obtained by the use of bifunctional ligands or by applying pressure on dried nanocrystalline matter, have been used for electrical transport measurements [686–689]. Pellets of Au_{55} and Pd_{561} nanocrystals exhibit nonmetallic behavior with specific conductivities in the range of $10^6\,\Omega^{-1}\text{cm}^{-1}$ [686–688]. The conductivity, however, increases dramatically with an increase in the diameter of the nanocrystals. An insulator–metal transition has been observed in pellets of ~12.5 nm Au and Ag nanocrystals [689]. Tunable transport phenomena in arrays of PbSe nanocrystals have been investigated [690]. The system evolves from an insulating regime dominated by Coulomb blockade to a semiconducting regime, where dominant transport mechanism is by hopping conduction.

Electrical transport measurements on layer-by-layer assemblies of nanocrystals on conducting substrates have been carried out with a sandwich configuration [691–693]. Nanocrystalline films with bulk metallic conductivity have been realized with Au nanocrystals of 5 and 11 nm diameter spaced with ionic and covalent spacers [692, 693]. The conductivity of monolayered two-dimensional arrays of metal nanocrystals has been examined with patterned electrodes [694–699]. Structural disorder and interparticle separation distance are found to be key factors that determine the conductivity of such layers [694–697]. The conductivity of the layers can be enhanced by replacing the alkane thiol with an aromatic thiol in situ [698,699]. The interaction energy of nanocrystals in such organizations can be continually varied by changing the interparticle distance.

Heath and coworkers [700, 701] prepared a monolayer of Ag (~3 nm) nanocrystals at the air–water interface in a LB trough and varied the interparticle distance by applying pressure. A host of measurements including reflectivity and nonlinear optical spectroscopy carried out in situ revealed the occurrence of a reversible Mott–Hubbard metal–insulator transition in the nanocrystal ensemble wherein the Coulomb gap closes at a critical distance between the particles. Tunneling spectroscopic measurements on films of 2.6 nm Ag nanocrystals capped with decanethiol show a Coulomb blockade behavior attributable to isolated nanocrystals [701]. On the other hand, nanocrystals capped with hexane and pentane thiol exhibit characteristics of strong interparticle quantum mechanical exchange (see Fig. 4.10). A similar behavior is observed in the case of self-assembled two-dimensional arrays of Co and Au nanocrystals [702, 703]. By varying the temperature, Rao and coworkers [431] have been able to tune the specific resistivity of a thin film of Au nanocrystals prepared at the interface of water and toluene in the range of tens of MΩ to a few Ω (see Fig. 4.11). Along with the change in conductivity, the nature of

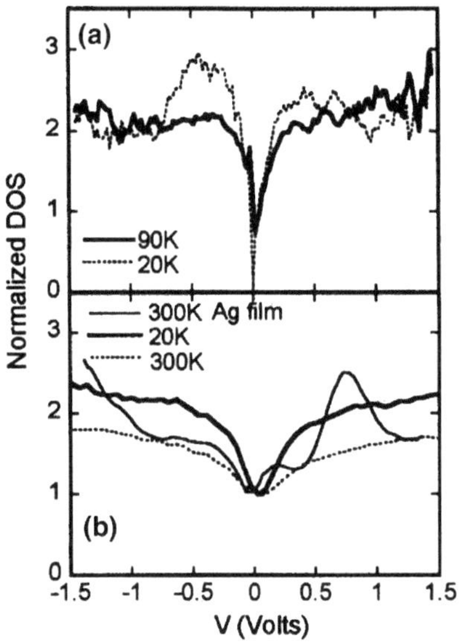

Fig. 4.10. Normalized density of states (DOS) measured from arrays of Ag nanocrystals of diameter ~2.6 nm capped (**a**) decanethiol and (**b**) hexanethiol at various temperatures. The temperature dependence of DOS near 0 V for decanethiol capped particles indicates that the films are nonmetallic. In the case of hexanethiol capped nanocrystals, the DOS around 0 V is temperature independent revealing the metallic nature of the film (reproduced with permission from [700])

the film also undergoes a change from that of an insulator to a metal. Schmid and coworkers [704] as well as Pal et al. [705] have obtained rectifying behavior from ensembles of nanocrystals. Although a clear understanding is yet to emerge, it is believed that the observed rectification is mainly due to the high charging energies of the individual nanocrystals.

4.2.1 Catalysis and Reactivity

Metal nanocrystals have been used as catalysts in commercial processes. The use of fine particles is attractive in view of their high surface areas per unit volume. The catalytic activity of nanocrystals is affected by other factors as well. A brief illustration is provided by the findings on Au catalysis. Despite its reputation as a noble metal, Au is found to be catalytically active at the nanoscale. An early study by Haruta et al. [706] found that Au nanocrystals embedded in oxide supports such as α-Fe_2O_3, Co_3O_4, and NiO were highly active for CO oxidation even at temperature as low as 200 K. Au nanocrystals with diameters in the range of 5–10 nm supported on γ-Al_2O_3 were capable of catalyzing CO oxidation [707]. Larger Au nanocrystals were not catalytically active. Goodman and coworkers [708] have observed that Au nanocrystals supported on titania exhibit a marked size effect in their catalytic ability for

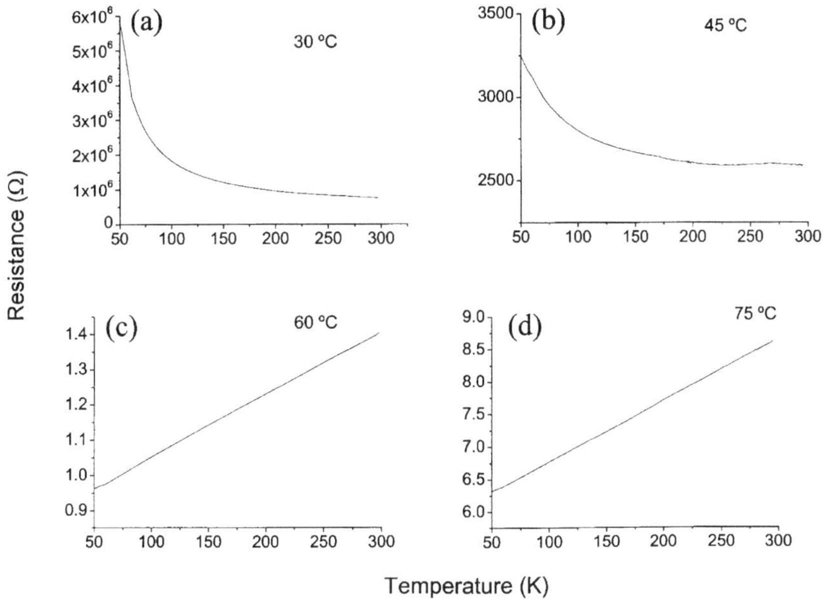

Fig. 4.11. The resistance of Au nanocrystal films as a function of temperature. The films (**a**)–(**d**) were obtained by carrying out the synthesis at different temperatures (**a**) 30°C, (**b**) 45°C, (**c**) 60°C, and (**d**) 75°C (reproduced with permission from [431])

CO oxidation, with Au nanocrystals in the range of 3.5 nm exhibiting maximum chemical reactivity. Tunneling spectroscopic measurements reveal that accompanying maximum catalytic activity is a metal to nonmetal transition, as the cluster size is decreased below 3.5 nm^3. In another study, small Au particles (<5 nm) supported on a zinc oxide surface exhibited a marked tendency to adsorb CO while those with diameters above 10 nm did not significantly adsorb CO (see Fig. 4.12) [709]. The increased activity of the Au particles is attributed to the charge transfer between the oxide support and the particle surface. The effect of the nanoparticle shape on the catalytic properties has been investigated. Tetrahedral Pt nanoparticles are better catalysts than spherical Pt nanoparticles for the Suzuki reaction between phenylboronic acid and iodobenzene [710]. Core–shell nanoparticles have also been tested for catalytic activity. Thus, Au–Pd nanocrystals supported on TiO_2 appear to have a high turnover frequency for the oxidation of alcohols [711]. The nanocrystals consist of Au-rich cores and Pd-rich shells indicating that the Au electronically influences the catalytic properties of Pd.

Alivisatos and coworkers have synthesized hollow nanocrystals by a process analogous to the Kirkendal effect observed in the bulk [712]. In bulk matter, pores are formed in alloying or oxidation reactions due to large differences in the solid-state diffusion rates of the constituents. By reacting Co nanocrystals

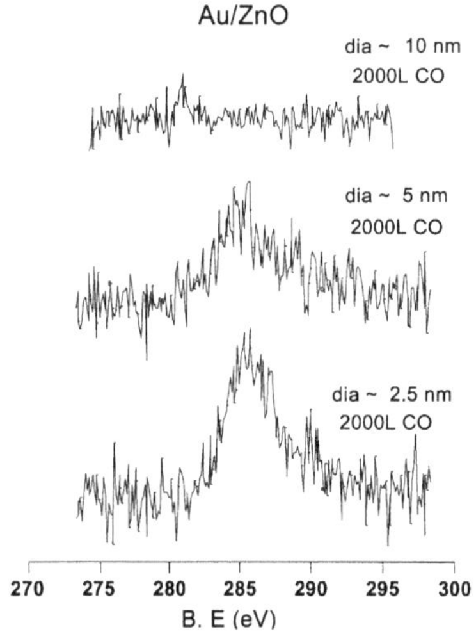

Fig. 4.12. C(1s) core-level spectra of CO adsorbed on Au particles supported on a ZnO substrate. The feature at 285 eV corresponds to molecularly absorbed CO. The diameters have been obtained from the metal coverage (reproduced with permission from [709])

with S, Se, or oxygen, hollow Co nanocrystals have been obtained. By starting with core–shell nanocrystals of the form Pt–Co and carrying out the process of hole creation, egg-yolk like nanostructures consisting of Pt yolk like core and Co oxide shell have been obtained.

4.3 Optical Properties

Optical properties of nanocrystals have been of interest for centuries (see Sect. 1.1) and have become the subject matter of several books in recent years [50, 73–75] and reviews [66–69, 713]. The plasmon resonance band has emerged as a probe of events taking place in the proximity of metal nanocrystals. Advances have been made in understanding the electronic structure of semiconductor nanocrystals from the excitonic absorption spectra.

Optical properties of nanocrystals of metals such as Au across a large size range reflect the changes in the electronic properties that occur with the variation in size. While it is ideal to study the optical properties using nanocrystals prepared by a single method, no single method, chemical or physical, yields nanocrystals of all the required size ranges. However, some general observations may be made. The normal position of the plasmon band of Au is around 520 nm. If the size of the Au nanocrystals is varied in range of 20–80 nm, the plasmon band broadens and shifts to the red with increase in size (see Fig. 1.16). This shift is due to the increased contribution of the higher order modes to the plasmon resonance band. Extinction due to scatter also plays a significant role in broadening the band. A different trend is apparent in the size range of 2–20 nm, a size range that is accessible by several synthetic schemes. The position of the band in this regime is nearly size-independent and Mie's theory suffices to explain the features. In this size regime, the experimental observations can be deduced by an examination of (1.17). The absorption is dominated by dipolar excitations, and with the decrease in size, the discreteness of the energy bands causes a fall in the free carrier concentration. As a consequence, the plasmon band gets dampened and broadened and eventually disappears. This is a manifestation of the size-induced metal–insulator transition. Mie's theory based on classical electrostatics is not valid in this size range, but, given the success at other size ranges, it has been possible to make the theory fit experimental observations by altering the mean free path of electrons (see (1.20)) [75]. Using Mie's theory and the discrete dipole approximation, absorption and scattering efficiencies and optical resonance wavelengths for Au nanospheres, SiO_2–Au nanoshells, and Au nanorods have been calculated. Au nanospheres show absorption and scattering cross-sections several orders higher than the conventional dyes. Systematic trends in such properties are of use in biological and cell imaging [714]. There have been attempts to incorporate Kubo's theory into Mie's theory [69, 66]. The plasmon band of metal nanocrystals is sensitive to the refractive index of the medium, as illustrated by the change in colors of Au nanocrystals embedded in SiO_2–TiO_2 gel films with the refractive index varying between 1.41 and 1.94 [715]. Plasmon bands of nanocrystals of Au–Ag, Au–Cu, and Au–Pt alloys have been studied in silica gel matrices [716,717]. The shift in the plasmon band due to alloying has also been investigated [432].

Just as Au nanocrystals, metallic nanoparticles of ReO_3 (Note ReO_3 looks like copper and conducts like copper) also exhibit a plasmon band centered around 520 nm [250]. Size-dependent changes such as a red shift of absorption band with increase in size are seen in these nanocrystals as well (see Fig. 4.13).

Kerker and Henglein have carried out detailed studies of the changes in the plasmon resonance band accompanying chemisorption, ligand desorption, and similar reactions [713, 718–721]. For example, adsorption of phosphine on Ag nanocrystals (6 nm) results in a blue shift of the plasmon band accompanied by broadening. A change in the refractive index of the solvent from 1.33 to 1.55 causes a red shift of the absorption maximum by 8 nm in the case of alkanethiolate-protected Au nanocrystals (diameter, 5.2 nm) [76]. The rate of flocculation brought about by aliphatic thiolacids attached to Au nanocrystals has been estimated based on the increase in the long wavelength scatter [722]. Metal ions binding to lipoic acid (a thiol acid) attached to Au or Ag nanocrystals are shown to bring about a strong dampening of the plasmon band (see Fig. 4.14), the extent of dampening depending on the nature of the metal ion [723]. The use of large metal nanocrystals results in more dramatic changes. Kim et al. found that heavy metal ions binding to Au nanocrystals

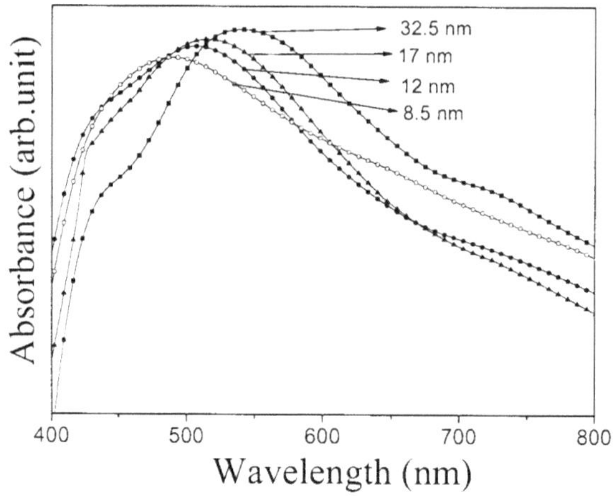

Fig. 4.13. Electronic absorption spectra of ReO_3 nanoparticles of different diameters (reproduced with permission from [250])

of diameters larger than 10 nm, induce a strong dampening of the absorption and have suggested that the process could be used to sense the presence of heavy metal ions which otherwise do not have a signature in the visible spectrum [724].

Significant difficulties are encountered in forming stable dispersions of large nanocrystals and a number of studies have, therefore, been carried out in the solid state. Optical properties of star-shaped Au nanoparticles, in particular plasmon resonance, have been examined [725]. Cu nanocrystals obtained by the reverse micelle method show optical properties dependent on the relative proportions of spheres and nanodisks [726]. Employing lithographically fabricated triangular Ag nanocrystals on glass or mica substrates, van Duyne and coworkers [727, 728] have found that the solvent refractive index can result in shifts of up to 100 nm. Adsorption of alkane thiol on these nanocrystals results in a blue shift (of up to 40 nm) in the absorption band depending on the length of the alkyl chains. It is estimated that a shift of 3 nm occurs for an increase of the chain length by a methylene unit. Such changes in the plasmon band of large Au nanoparticles are reproducible and have prompted commercial enterprises like Biacore to manufacture a kit to detect the binding of large biomolecules such as enzymes and antibodies. Laurent et al. [729] have shown that surface plasmons of Au nanoparticle arrays can be imaged by far-field Raman spectroscopy.

Colloid scientists have for long been seeking ways to study the optical properties of low-dimensional aggregates of nanocrystals. Quinten and Kriebig [730] have calculated the optical properties of gold nanoparticles with diameters of 10 and 56 nm arranged in small aggregates with varying shapes and sizes. They have shown that additional bands at long wavelengths occur if the distance between the flocculating spheres is small relative to the particle

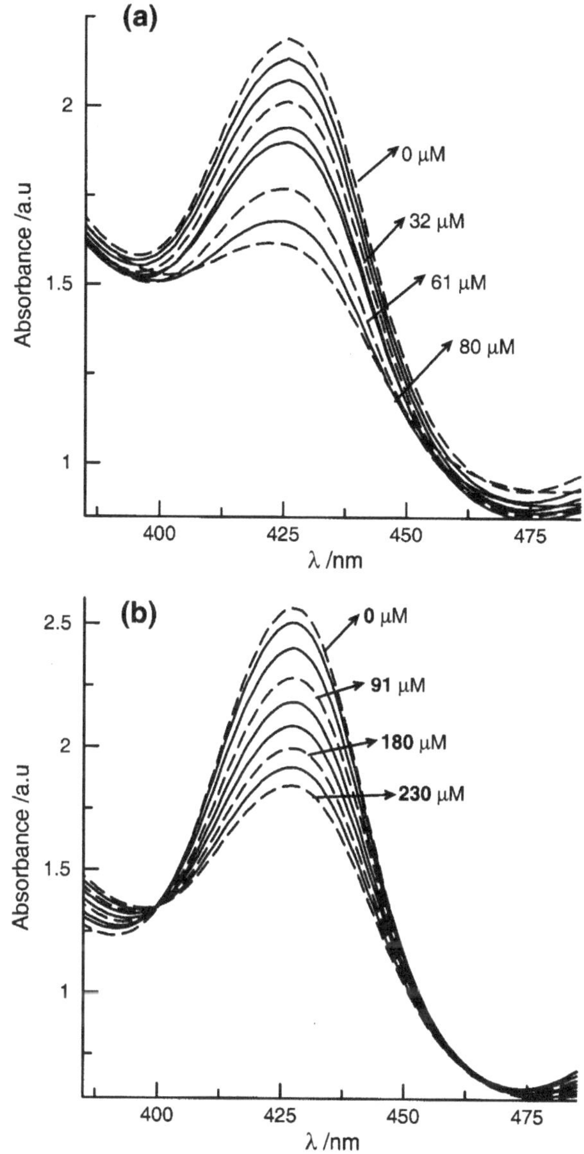

Fig. 4.14. Electronic absorption spectra of ~5 nm Ag nanoparticles showing changes accompanying the addition of (**a**) Cu^{2+} and (**b**) Fe^{2+} ions. The concentrations of the ions are indicated (reproduced with permission from [723])

radius. By using of lithographically fabricated Au nanocrystal pairs, Leitner and coworkers [731] have measured the changes in the optical properties of a nanocrystal induced by a neighbor. Maier et al. [732] have investigated the optical properties of linear rows of Au nanocrystals and found a strong red shift in the plasmon resonance band when light polarized along the direction of the row was incident. Dielectric-metal hybrid structures resembling a grain of rice have been designed and fabricated, combining the intense local fields of nanorods and the tunable plasmon resonance of nanoshells [733].

Excitonic absorption of semiconductor particles is described by the effective mass approximation [78, 753]. The optical properties of the particles are understood in terms of three distinct regimes. The weak confinement regime is one where the exciton diameter (α_b) is less than R. A typical example being CuCl nanocrystals ($\alpha_b = 0.7\,\text{nm}$) with R in the range of few nanometers. The intermediate confinement regime comprises all sizes where R is comparable to α_b. In this regime, it is plausible that R falls in the region between the Bohr radii of the electron and the hole (α_e) leading to interesting consequences. Spectral characteristics indeed depend on the nature of confinement of the charge carriers. In the strong confinement regime, R is less than α_b. In the weak confinement regime, the absorption spectrum shifts to higher energies with decrease in the size of the nanocrystal. The energy corresponding to the onset of absorption is proportional to $1/R^2$. Thus, we have

$$\Delta E = \frac{\hbar^2\pi^2}{2R^2}\left[\frac{1}{m_e^*} + \frac{1}{m_h^*}\right] - \frac{1.786e^2}{\epsilon R} - 0.248E_{Ry}^* \tag{4.5}$$

where E_{Ry}^* is

$$\frac{e^4}{2\epsilon^2\hbar^2(m_e^{*-1} + m_h^{*-1})} \tag{4.6}$$

Such size dependence has been observed in nanocrystals of I–VII and II–VI compounds such as AgI, CuCl, and CuBr [50, 78, 754, 755]. The bandgap of these nanocrystals could be ascertained from the absorption onset as shown in Fig. 4.15 [756,757]. A similar dependence of the absorption onset is seen in the other two regimes as well. However, other spectral characteristics are affected differently depending on whether the electrons or the holes suffer a strong or intermediate confinement. In the intermediate confinement regime, the spectra in the high wavelength region resolve into bands. If $\alpha_e > R > \alpha_h$, donor-like exciton dominated by hole motion is seen. If $\alpha_h > R > \alpha_e$, an acceptor-like exciton dominated by electron motion is seen. Evidence for a donor-like exciton has been obtained in the case of large CdS and CuBr nanocrystals [758–760]. An acceptor-like exciton is seen in the case of PbI_2 nanocrystals [761]. In the strong confinement regime, the spectrum is reduced to a series of wide bands corresponding to the transitions between electron and hole levels.

Optical properties of most known semiconductors have been investigated [762] and the effective mass approximation is satisfactory for describing the changes in the optical properties of semiconductor nanocrystals in the weak confinement regime. In the intermediate and strong confinement regimes, significant deviations are seen. Tight-binding approximation has been used to

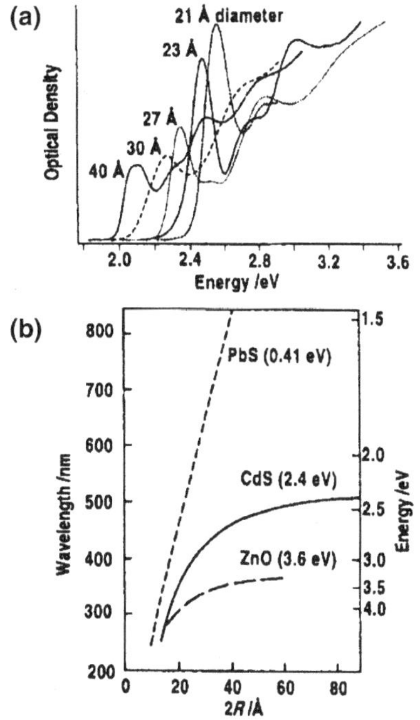

Fig. 4.15. (**a**) Absorption spectra of CdSe nanocrystals (at 10 K) of various diameters. (**b**) Wavelength of the absorption threshold and band gap as a function of the particle diameter for various semiconductors. The corresponding energy gap in the bulk state is given in parenthesis (reproduced with permission from [662])

describe the properties of nanocrystals in the strong and intermediate confinement regimes [81, 763–766]. The tight-binding parameters, optimized to reproduce the experimentally obtained bulk band gaps, are applied to nanocrystals to obtain the band structure, assuming that they possess the same structure. Tight-binding methods provide reliable descriptions of the experimental data, but, due to the very nature of the method, it is not possible to derive a general equation to describe size-dependent changes. The changes have to be calculated individually for each system.

Electronic absorption spectra of CdS nanocrystals over an extended size range is shown in Fig. 4.16. The onset of the absorption band which is indicative of the band gap, shifts from the bulk value of ~495 nm(2.5 eV) well into the UV region (4.0 eV, ~300 nm) as the diameter of the nanocrystals changes from 9.6 to 1.28 nm. There has been considerable experimental and theoretical effort to understand size-dependent changes of the optical spectra of CdS and other chalcogenide nanocrystals. In Fig. 4.17, the results of the theoretical calculations and the experimentally observed changes in the band gap are compared. Tight binding approximation does indeed yield better results.

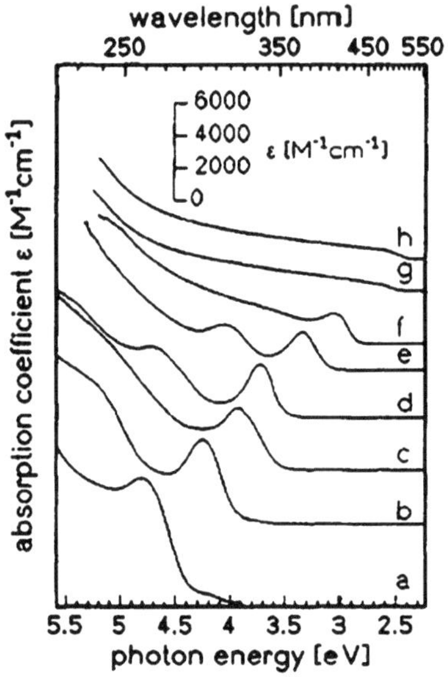

Fig. 4.16. Absorption spectra of CdS nanocrystals of different sizes prepared in aqueous medium. The diameters are (**a**) 1.28 nm, (**b**) 1.44 nm, (**c**) 1.60 nm, (**d**) 1.86 nm, (**e**) 2.32 nm, (**f**) 3.88 nm, (**g**) 5.6 nm, and (**h**) 9.6 nm (reproduced with permission from [269])

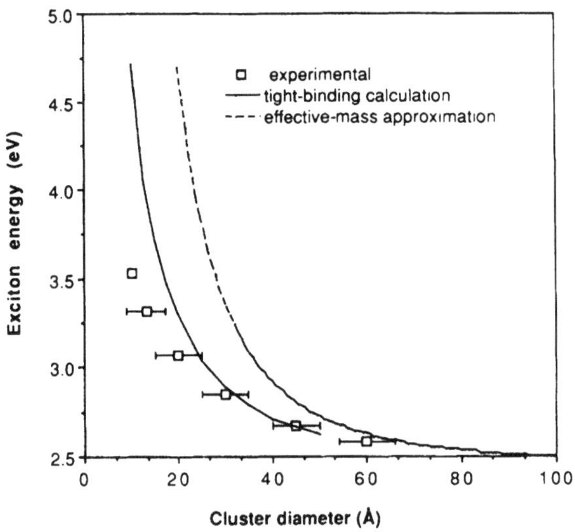

Fig. 4.17. Variation of the bandgap (exiton energy) of CdS nanocrystals as a function of size compared with results from effective mass and tight binding approximations (reproduced with permission from [763])

Nanocrystals of semiconducting oxides such as ZnO possess certain advantages compared to other semiconducting materials. Semiconducting oxides are not air sensitive, less toxic, and are transparent. For these reasons, nanocrystalline semiconducting oxides have been well studied. A wide range of Bohr radii and electron and hole effective masses have been reported for ZnO nanocrystals. Quantum confinement effects in wurtzitic ZnO nanocrystals manifest themselves below the diameter of 7.0 nm [766–768]. Absorption spectra of ZnO nanocrystals of different diameters are shown in Fig. 4.18. ZnO nanocrystals exhibit an increase in the band-gap with decrease in diameter much like CdS. Besides establishing the size-dependent UV absorption spectra of ZnO nanocrystals, the spectra of Mn-doped ZnO nanoparticles have been studied [734]. Such doping affects the band gap of ZnO. The UV band intensity of ZnO nanoparticles have been employed to investigate the growth process of capped ZnO nanocrystals [735]. The growth is nonmonotonic, but different from that found in the growth of Au nanocrystals [736]. ZnO Carrier recombination process in quantum dots (4 nm diameter), nanocrystals (20 nm diameter), and bulk ZnO have been investigated by PL spectroscopy in the 8.5–300 K range. Recombination of the acceptor bound excitons cause PL in quantum dots and nanocrystals, especially at low temperatures [737]. Band-edge luminescence in octylamine-capped ZnO nanocrystals has been measured and size dependent band gaps are estimated and compared with the values from theoretical models [738].

Other semiconductor systems in which strong confinement have been obtained include the sulphide and selenides of lead, nitrides of Ga and In. The band-gap of hexagonal InN has been recently shown to be around 0.7 eV, markedly lower than the previously accepted value of 1.9 eV. Rao and

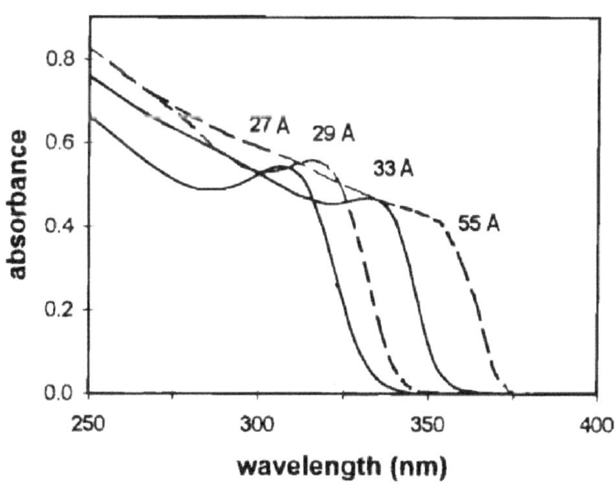

Fig. 4.18. Absorption spectra of ZnO nanocrystals of different diameters. The diameters are indicated (reproduced with permission from [768])

coworkers [249] have been successful in establishing that nanocrystals and other nanostructures of InN do indeed posses a band gap of 0.7 eV. The absorption onset of PbSe nanocrystals shifts from 2,200 to 1,200 nm as the diameter is varied from 9.0 to 3.0 nm [334]. Interestingly, the absorption spectra of PbS nanocrystals can be modified by the application of electric fields [739]. The valance band maxima and absorption bands of CdTe nanoparticles blue shift as the size decreases due to quantum confinement [740]. The band gap of CdTe nanocrystals can be engineered through surface modification by thiolate ligands under ambient conditions [741].

Electronic properties of semiconductor ensembles have been examined by absorption spectroscopy [770]. Interparticle interactions could lead to a discernable change in the optical properties of a lattice of semiconductor nanocrystals. The coupling of the electronic states could vary from weak to strong. Absorption spectra showing the results of weak coupling in CdSe nanocrystals are shown in Fig. 4.19. When present in close-packed organization, the absorption spectra of the nanocrystals are broadened and red-shifted.

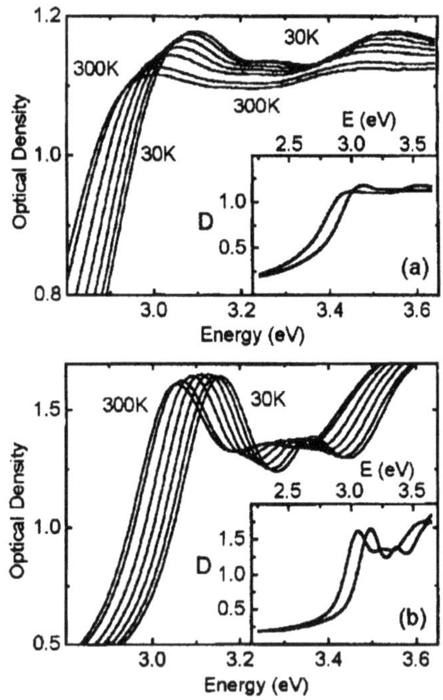

Fig. 4.19. Absorption spectra of thin films of close-packed (**a**) and isolated (**b**) CdSe nanocrystals at different temperatures (from *right* to *left curves*); 30, 80, 130, 180, 230, 280, and 300 K. The insets show the full-range spectrum of optical density D for lowest and highest temperatures. A red shift and broadening of the peaks is seen in the case of close-packed films (reproduced with permission from [771])

This change is attributed to interparticle dipolar interactions [771]. Bawendi and coworkers [772] have studied the changes in ensembles of CdSe nanocrystals of different diameters and obtained evidence for long-range resonance transfer of electronic excitation from smaller to bigger nanocrystals due to dipolar interactions.

In a noteworthy experiment, Weller and coworkers [773] prepared drop-cast films of giant CdS clusters of the type $Cd_{17}S_4(SCH_2CH_2OH)_{26}$ and $Cd_{32}S_{14}(SCH_2CH(CH_3)OH)_{36}$ with diameters of 1.4 and 1.8 nm, respectively and studied their optical properties. An integrating sphere was used to collect absorption data, thereby virtually eliminating errors from inhomogenities and size distributions. Reflection spectra of micron-size crystals of $Cd_{17}S_4(SCH_2CH_2OH)_{26}$ show red-shifts of the absorption onset as well as the broadening of the excitonic peak (see Fig. 4.20) [774]. The experiments of Weller support the idea of dipolar interaction leading to the red-shift and broadening. The signature of such interactions has been observed in CdS multilayer deposits as well [775]. Exchange interactions arising from strong coupling of the electronic states could lead to large scale delocalization of electronic states. Such delocalization has been observed in CdSe nanocrystals. Gaponenko and coworkers [776] have shown that the optical properties of an ensemble of small (1.6 nm) CdSe nanocrystals are similar to those of bulk CdSe, due to complete delocalization of the electronic states of individual nanocrystals.

Luminescence properties of semiconducting nanocrystals have been a subject of considerable attention in the past few years. It has been realized that

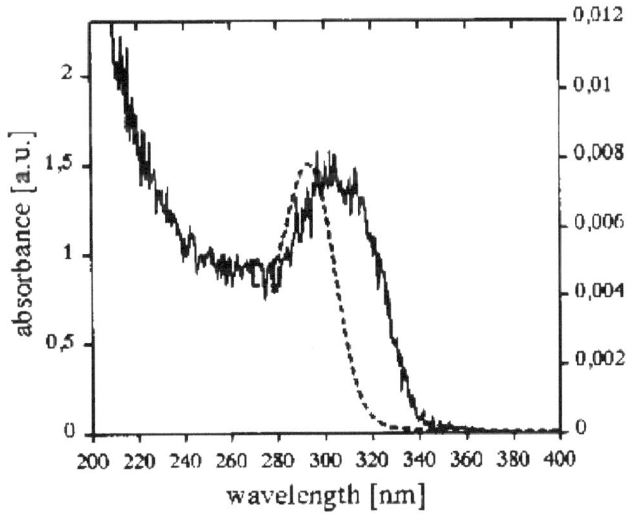

Fig. 4.20. Absorption spectrum of $Cd_{17}S_4(SCH_2CH_2OH)_{26}$ nanocrystals dispersed in solution (*dotted line*) and reflection spectrum of micron-sized crystals of the same compound (*solid line*) (reproduced with permission from [774])

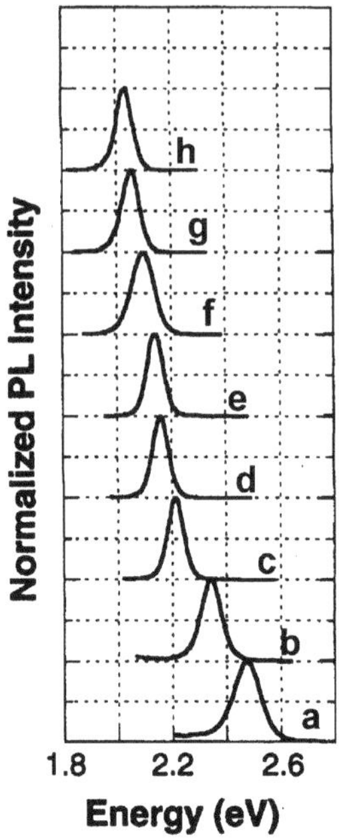

Fig. 4.21. The emission spectra of CdSe nanocrystals of different sizes (**a**) 2.4 nm; (**b**) 2.5 nm; (**c**) 2.9 nm; (**d**) 3.3 nm; (**e**) 3.9 nm; (**f**) 4.1 nm; (**g**) 4.2 nm; and (**h**) 4.4 nm. Broader emission peaks at small sizes result due to wider size distributions of the nanocrystals (reproduced with permission from [777])

intense and narrow emission can be brought about by changing the size of the nanocrystals [777]. The emission spectra of CdSe nanocrystals of different sizes is shown in Fig. 4.21. The emission band enables a study of quantum-confinement effects. In the case of GaN nanocrystals, a size-dependent band at ~315 nm is shown to be truly indicative of quantum-confinement. This occurs in GaN nanocrystals with diameters in the range of 2.5–10.0 nm (see Fig. 4.22) [247]. However, in most cases, the as-synthesized semiconductor nanocrystals emit poorly, due to the mid-gap surface states created by surface atoms and defects. For example, a broad emission peak stretching across the entire visible region is seen in the case of ZnO nanocrystals due to defects in the lattice [767,768]. The surface states interfere with the decay process and result in poor quantum yields. The surface states could be lifted out of the midgap region by treating the surface of the nanocrystals with appropriate capping agents. Another trick adopted is to grow a layer of wider band gap material (preferably epitaxially). Such changes are also brought about by doping the nanocrystal with other ions. $CuInS_2$ Nanocrystals obtained by heating an organometallic precursor offer tunability of optical properties. Introduction of Zn increases the PL intensity [769].

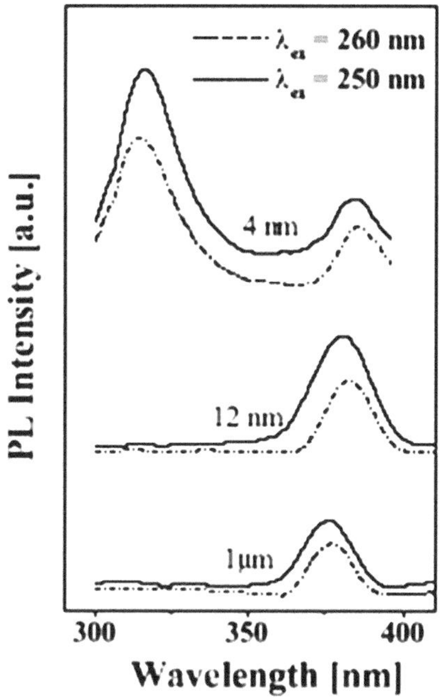

Fig. 4.22. Emission spectra of GaN nanocrystals of different diameters. The diameters are indicated (reproduced with permission from [247])

Fluorescence from semiconductor nanocrystals are superior to conventional dyes in several respects. The emission peaks are intense, narrow, and are tunable. Further, Stokes shifts much larger than those of conventional dyes are possible. Employing nanocrystals of different sizes, but of the same material, emission at widely different wavelengths can be brought about by excitation at a single wavelength. The earlier factors have led to the exploration of a wide range of applications for luminescent semiconductor nanocrystals. For example, CdSe nanocrystals with green and red emission (of ~3 and 5 nm diameter, respectively) can be used to produce white light by mixing them with the blue emission from an LED. White emission is also observed from extremely small CdSe nanocrystals [748]. These nanocrystals are presumed to have the structure of highly stable magic number CdSe clusters observed in mass spectrometry [749]. CdSe–CdZnS core–shell nanocrystals exhibit multiphoton emission [750], while CdSe–ZnO nanoparticles are directional photoemitters with tunable wavelength [751]. Silicon quantum dots obtained by plasma synthesis and subjected to wet chemical surface passivation with organic ligands, exhibit photoluminescence quantum yields in excess of 60% at ~790 nm [752]. Such high yields are surprising since the emission in the case of indirect band gap materials such as Si, is entirely defect-mediated and is not related to the band gap. The emission band of nanophospor material is sensitive to the refractive index of the material. There has been considerable work on nanophosphors because of their potential application in displays and devices. Chander [742] has recently reviewed nanophosphors.

The discovery of the phenomenon of multiple exciton generation in quantum dots of PbS and PbSe has generated considerable excitement [743–745]. It seems possible that lead chalcogenide quantum dots when irradiated with photons of four times the band gap energy, result in the generation of up to three excitons possessing the band gap energy. In other words, quantum yields of the order of 200–300% have been observed. The experimental results confirm the earlier theoretical prediction of Nozik [746, 747]. These observations, though indirect, have implications for applications such as solar power generation.

4.4 Magnetic Properties

Magnetic properties at the nanoscale strongly depend on the interplay involving the surface atoms and the energetics of spin reversal. When particles of a wide range of sizes are investigated, their coercivity tends to follow a trend illustrated in Fig. 1.8 [41, 42]. For example, the coercivity of Fe nanocrystals embedded in a Mg matrix increases in the range of 4–20 nm and decreases with further increase in size [778]. Thus, maximum coercivity is achieved when the particle size is 20 nm. In simple terms, higher than bulk coercivity is obtainable in single domain particles as magnetization reversal occurs by coherent rotation instead of domain wall motion, the common cause for coercivity of multidomain particles of larger sizes. As the size is further lowered, thermal reversal of spins especially of the surface atoms becomes increasingly easy, leading to a loss of coercivity and eventually to superparamagnetism. Changes in coercivity with size can often be complicated. In the nanoregime, it is possible that shape anisotropy has a higher value than the magnetocrystalline anisotropy, specially in elongated particles and nanorods. In the case of spherical particles, surface anisotropy could dominate over magnetocrystalline anisotropy. Thus, γ-Fe_2O_3 nanocrystals undergo a change of anisotropy from cubic to uniaxial as the size of the nanocrystals is reduced below 11 nm [779, 780]. Bean and coworkers [781, 782] have proposed a type of magnetic anisotropy called exchange anisotropy in small electrodeposited Co nanoparticles surrounded by CoO layers. Coupling between the ferromagnetic Co and the antiferromagnetic CoO is such that the reversal spins of Co core is resisted by the CoO shell, resulting in higher coercivity values. Such a behavior is also seen in Fe nanoparticles [784]. Skumryev et al. [785] suggest that the superparamagnetic limit can be beaten by the use of exchange bias. They have succeded in making Co–CoO nanocrystals as small as 4 nm, ferromagnetic at room temperature. In general, as the size decreases, coercivity is lowered. However, some theoretical studies suggest that coercivity should actually increase with the decrease in particle size if surface anisotropy effects are taken into account [783].

The loss of coercivity with size, i.e., the superparamagnetic nature of nanoparticles, has been well studied. The discussion earlier deals with the behavior below the blocking temperature. Neel's theory suggests an exponential behavior for the temperature induced relaxation (τ):

$$\tau = \tau_0 e^{KV/k_b T}, \quad (4.7)$$

where K is the anisotropy constant and V the particle volume. Using realistic values for τ_0 and the timescale of a typical magnetometer, one arrives at the relation,

$$T_b = \frac{KV}{25k_b}, \quad (4.8)$$

where T_b is the blocking temperature [786–788]. In Mössbauer spectra, a magnetic sample exhibits a sextet of lines below the blocking temperature and a doublet above it (see Fig. 4.23). The blocking temperature measured using Mössbauer spectroscopy is related to T_b by the relation,

$$T_b(Mössbauer) \simeq 5.5T_b. \quad (4.9)$$

By measuring T_b using Mössbauer spectroscopy and magnetic measurements, one obtains estimates of the particle size as well as the magnetic anisotropic properties.

Much of our understanding of magnetic properties of fine particles is derived from studies on magnetic oxide systems, especially spinels. Spinel ferrites with the formula MFe_2O_4 (M=Mn, Mg, Zn, Co, Ni, Fe) with a face-centered cubic unit cell contain eight formula units (see Fig. 4.24). Two sites, the tetrahedral A site and the octahedral B site, are available for cation occupation. A variety of size-dependent changes are observed with spinel ferrite nanocrystals. For example, a blocking behavior has been observed in $CoFe_2O_4$ [790] and $MnFe_2O_4$ [791]. The expected fall in coercivity with size occurs in most spinel ferrite nanoparticles. Depending on the method of preparation, the nanoparticles exhibit different properties due to differences in the cation occupancy. Locations of the trivalent and bivalent cations affect the magnetic exchange interaction markedly, leading to differences in the magnetic properties. $CoFe_2O_4$ nanoparticles exhibit a blocking temperature 150 K higher than the Mn ferrite nanoparticles [788]. Tetragonal $CoMn_2O_4$ nanocrystals (5–12 nm diameter) prepared by the decomposition of Co–Mn acetylacetonate in oleylamine are ferromagnetic at low temperature and paramagnetic at room temperature, with a blocking temperature of 30–40 K. The particles also show an exchange bias behavior due to ferri-ferromagnetic coupling [789]. A size-dependent rise in T_c is observed for Mn ferrite nanoparticles [792, 793]. It is not clear if the cation redistribution or the finite size effect is responsible for this observation [794]. In addition to cation site disorder, surface disorder brought about by the curvature of nanoparticle surfaces influences the magnetic properties of nanocrystals of spinel ferrites and related oxide materials.

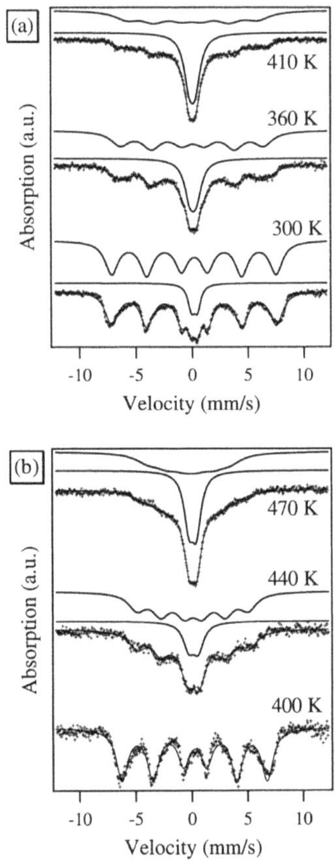

Fig. 4.23. Temperature dependent Mössbauer spectra of (**a**) 20 nm $MgFe_2O_4$ and (**b**) 20 nm $CoFe_2O_4$ nanoparticles (reproduced with permission from [788])

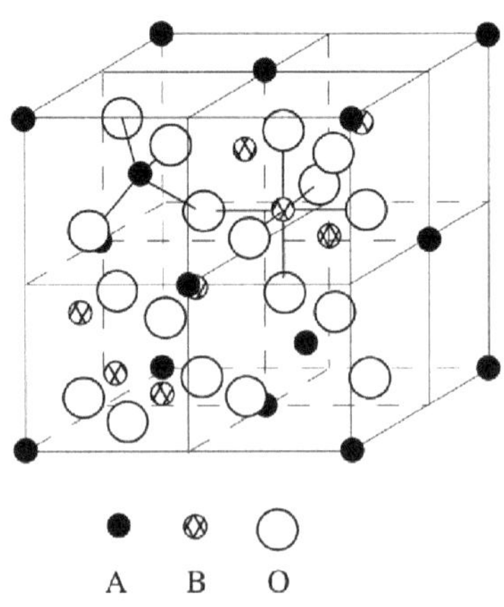

Fig. 4.24. Unit cell of spinel ferrites (reproduced with permission from [788])

Magnetic properties such as the saturation magnetization also exhibit size-dependent changes at the nanoscale. The saturation magnetization of nanocrystals is less than that of the corresponding bulk, because of the dead layer on the surface [795]. Figure 4.25 shows the variation of saturation magnetization with size for $MnFe_2O_4$ nanoparticles. A linear correlation is obtained with the inverse of diameters and can be useful to estimate the thickness of the magnetically inactive surface layer. Such a size-dependent reduction in saturation magnetization, first observed by Berkowitz and coworkers [796] in the late sixties, has been a subject of some debate. Both finite size effects [797,798] and surface spin disorder [799,798] have been proposed as likely causes. Recent studies indicate the latter to be the likely cause [795, 800, 801].

In certain cases, the surface layer makes a nanoparticle behave like a spin glass. Spin-glass behavior is observed in several nanoparticle oxide systems such as $NiFe_2O_4$ [802] and γ-Fe_2O_3 [803]. Marked changes in magnetic properties can be brought about by influencing the surface layer with the use of different capping agents [804].

A layer of uncompensated spins at the surface is responsible for the ferromagnetic interactions observed at low temperatures in normally antiferromagnetic oxides such as MnO, NiO, and CoO [44, 45, 805, 806]. Evidence for such a behavior can be seen from Fig. 1.10. In this study, the blocking temperature was shown to increase with increase in size for NiO and show the opposite trend for MnO nanoparticles. Nanoparticles of ReO_3 also turn ferromagnetic, exhibiting hysteresis at low temperatures (see Fig. 4.26) [250]. The

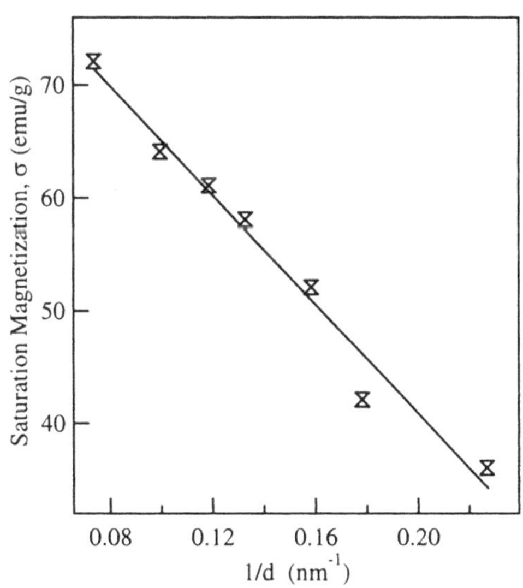

Fig. 4.25. Plot showing the variation of saturation magnetization with the inverse of mean diameter for $MnFe_2O_4$ nanoparticles at 20 K (reproduced with permission from [795])

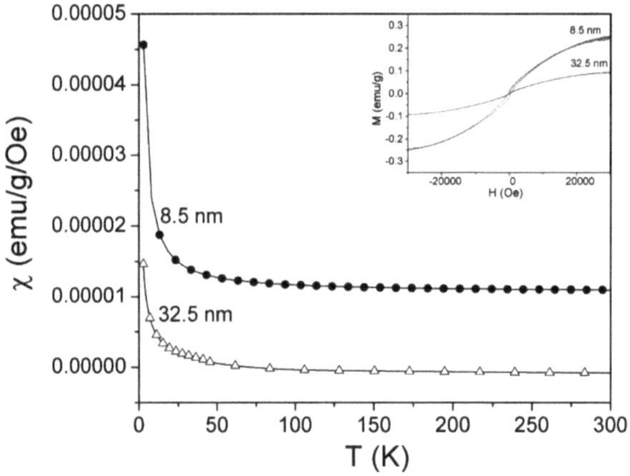

Fig. 4.26. Magnetic susceptibility vs. temperature curves of 8.5 and 32.5 nm ReO_3 nanoparticles. Inset shows the field dependence of magnetization of the 8.5 and 32.5 nm particles at 5 K (reproduced with permission from [250])

earlier observation is interesting since bulk ReO_3 is reported to be diamagnetic or paramagnetic.

It has been found recently that nanoparticles of all metal oxides, including those which are normally nonmagnetic such as Al_2O_3, SnO_2, and TiO_2, show ferromagnetic features in terms of magnetic hysteresis at room temperature due to surface effects [807].

Dilute magnetic semiconductors, typically transition metal ion doped chalcogenides and oxides have evoked keen interest in recent times because of their potential uses in areas such as spin-dependent electronics. The ability to synthesize quantum confined magnetic semiconductor nanocrystals represents a key challenge. Efforts to produce nanocrystalline Mn-doped II–VI magnetic quantum dots have yielded nanocrystals with widely different magnetic characteristics, depending on the nature of preparation and postsynthesis treatment such as thermal annealing. For example, Strouse and coworkers have found that Mn doped CdSe nanocrystals can be turned superparamagnetic by annealing the synthesized nanocrystals [808]. Unlike Mn-doped ZnO nanoparticles, Mn-doped GaN nanoparticles have been found to exhibit ferromagnetism [809].

Studies with cluster beams of magnetic materials have thrown light on the magnetic phenomena at the lower end of the nanoscale. The magnetic moments of size-selected 3d transition metal clusters have been measured using the Stern–Gerlach technique [810, 811]. An increase in the magnetic moment per atom with decreasing size is observed in Fe, Co, and Ni clusters (see Fig. 4.27). In simple terms, the average coordination of a surface atom is lowered resulting in an atom-like magnetic moment. Billas et al. [812] have provided a model assuming a spin glass-like behavior and the presence

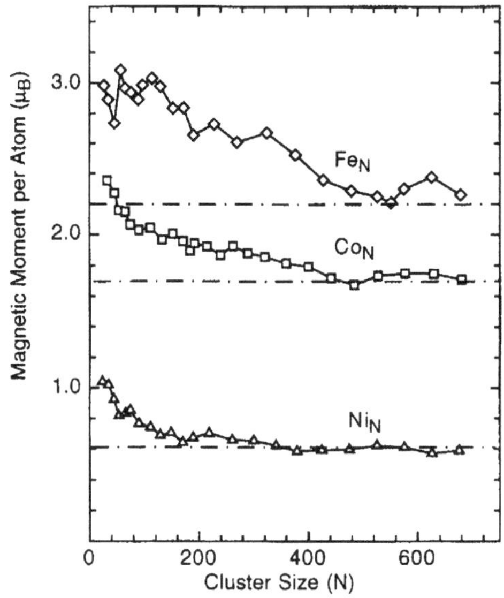

Fig. 4.27. Magnetic moment per atom for Fe, Co, and Ni clusters showing size-dependent changes (reproduced with permission from [811])

of magnetic shells to account for the fine structure in Fig. 4.27. A remarkable confluence of nanoscalar phenomena causes Rh clusters with nuclearities in the range of 9–31 to become ferromagnetic [813]. This observation was prompted by a theoretical investigation, attributing it to reduced coordination, lattice expansion and icosahedral symmetry of the clusters [814]. Further, magneto-transport measurements on Co nanoparticles in the single-electron tunneling regime show features of spin accumulation [815].

There have been attempts of varying degrees of sophistication to build solids of magnetic grains in which the interaction between grains is controllable. Such a scheme would eventually lead to a magnetic solid whose properties can be tailored to probe a phenomenon or to suit a particular application. Of particular interest are solids which consist of self-assembled nanocrystals which can exhibit collective properties. Nanocrystals of Co when organized into two-dimensional arrays exhibit a higher T_b compared to isolated nanocrystals, i.e., they display a higher resistance to thermal reversal of their spins than when they are isolated [162,702]. Sun et al. [297] have heat-treated a two-dimensional array of FePt alloy nanocrystals to cause phase segregation of Fe and Pt to obtain a ferromagnetic film made up of interacting nanocrystals. Such a film was capable of supporting high-density magnetic reversal transitions. Exchange spring magnet – nanocomposites consisting of magnetically hard and soft phases interacting via magnetic exchange coupling, have been made by annealing an ordered array FePt and Fe_3O_4 nanocrystals [816].

Pileni and coworkers [817–819] have carried out a series of experiments on drop-cast films of γ-Fe_2O_3 nanocrystals on graphite and have shown that the easy axis of these nanocrystals could be partially aligned parallel to substrate by the use of a magnetic field during evaporation of the nanocrystals. The alignment of the easy axis leads to a measurable anisotropy as seen in the hysteresis loops. By doping these nanocrystals with Co, the coercivity of such a film could be enhanced. On Si(100) substrates, the easy axis of magnetization of citrate-capped γ-Fe_2O_3 nanocrystals orients perpendicular to the substrate [820] giving rise to perpendicular magnetization.

5

Core–Shell Nanocrystals

Properties of a nanocrystal can be influenced markedly by encasing it in a sheath of another material [531]. The material of the shell in such a core–shell structure can be a metal, semiconductor, or an oxide. The shell material helps to impart novel, desired properties on the nanocrystals. For example, defects prevalent in the surface states of semiconductor nanocrystals can be transferred to a buffer layer of the shell material to obtain better emission from the nanocrystals. We use the notation, core–shell to denote core–shell structures. We employ the following classification to describe the nature of the core and the shell: semiconductor–semiconductor, metal–metal and metal–oxide, semiconductor–oxide and oxide–oxide. The classification is artificial in that the nanocrystals result from similar synthetic strategies. The motivation for carrying out the modification of the shell material, however, differs in each case.

5.1 Synthesis and Properties

5.1.1 Semiconductor–Semiconductor

Coating a semiconductor shell on a core semiconductor nanocrystal is usually carried out for one of the following reasons. The shell material provides a sink for the surface layer defects, thereby pushing the surface electronic states out of the midgap region. For this scheme to be successful, it is essential for the bandgap of the shell material to be higher than that of the core material. The second scheme is to coat a nanocrystal with a layer of narrower bandgap material that makes available an additional area for delocalization of electrons and holes. The top of the conduction band and the bottom of the valance band move closer, resulting in a lowering of the bandgap. Different synthetic schemes used to prepare nanocrystals are adapted to synthesize core–shell nanocrystals. A few key factors determine the synthetic scheme employed. An ideal method should yield only core–shell particles and not a mixture of particles. The growth of the shell layer should be uniform so that the core–shell nanocrystals are monodisperse. A reasonable lattice match is essential for epitaxial growth of the overlayer, else defects that develop the interface between the core and the shell layer hinder delocalization and would make the shell layer ineffective.

The inverted micelle method has been adapted to prepare CdSe–ZnSe [821], CdS–Ag_2S [822], CdS–CdSe, and CdSe–CdS nanocrystals [823, 824]. Growth in the form of core–shell structures is brought about by successive deposition of the core and shell materials in the water pool. For example, CdS–Ag_2S nanocrystals were obtained in two steps, by first preparing CdS nanocrystals by mixing together Cd^{2+}/AOT/heptane and Cd^{2+}/AOT/heptane micelles, and then growing the shell layer by injecting a $AgNO_3$ solution. The Ag^{2+} ions displace the Cd^{2+} ions from the surface [822]. The resulting core–shell nanocrystals exhibit a large increase in nonlinear absorption. In the case of CdSe–CdS, an increase in the emission intensity was seen [824]. CdSe–CdS core–shell nanoparticles with core diameter of 1.5 nm were obtained at the liquid–liquid interface starting with cadmium myristate and oleic acid in toluene and selenourea (thiourea) in the aqueous medium [825]. The method of Murray et al. involving the decomposition of dimethyl cadmium has been adapted to synthesize nanocrystals of CdSe–ZnS [83, 826], CdSe–ZnSe [827], and CdSe–CdS [828] (see Fig. 5.1). Core–shell growth is achieved in the above schemes by injecting the precursors forming the shell materials into a dispersion containing the core nanocrystals. The injection is carried out at a slightly lower temperature to force shell growth, avoiding independent nucleation. A mixture of diethylzinc and bis(trimethylsilyl)sulphide was injected into a hot solution containing the core CdSe nanocrystals to encase them with a ZnS layer [826]. The CdSe nanocrystals were grown at temperatures in the range of 300–340°C, while the shell material were grown at a temperature range of 140–220°C. Alivisatos and coworkers [828] have shown that an intermediate step of refluxing the nanocrystals in pyridine helps to detach the ligand shell and promotes epitaxial growth of the shell layer. The epitaxial nature of the shell has been confirmed by HRTEM and other measurements [828].

For a given nanocrystal core, the position and the intensity of an emission band depends on the thickness of the overlayer (see Fig. 5.2). The core–shell nanocrystals obtained by thermal decomposition methods are monodisperse and exhibit high quantum yields. In fact, they are among the brightest emitters known, with quantum yields of up to 90% in some cases. In other words, they are capable of emitting 20 times more light than a traditional organic dye. It is believed that the high temperature conditions during the synthesis promote defect-free growth and epitaxial overlayer formation. Similarly, doping nanocrystals may enhance their optical properties. Mn-doped CdS–CdS or ZnS nanoparticles show improved PL and may find application in displays and data storage [829]. O'Brien and coworkers [830, 831] have pioneered single-source methods to prepare core–shell nanocrystals. Thus, by successive thermolysis of unsymmetrical diseleno and dithio carbamates – $Cd(Se_2CNMe(Hex))_2$ and $Cd(S_2CNMe(Hex))_2$ – core–shell nanocrystals of

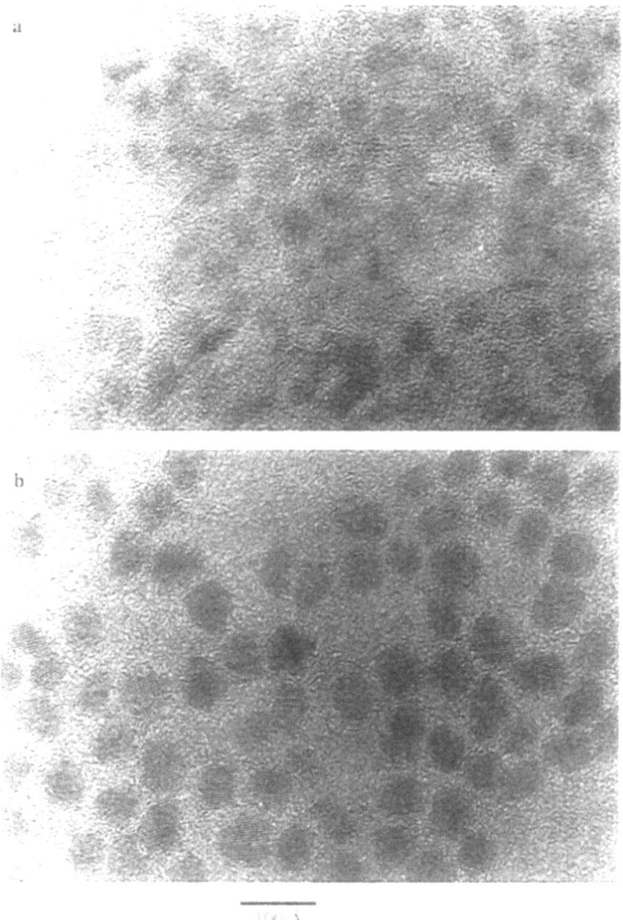

Fig. 5.1. HRTEM images of (**a**) CdSe core and (**b**) CdSe–CdS nanocrystals synthesized from the cores. The pictures are at the same magnification. The core diameter is 3.4 nm and the shell thickness for the core–shell nanocrystals is 0.9 nm. The nanocrystals in (**b**) are clearly larger than the ones in (**a**). Where lattice fringes are observed in (**b**), they persist throughout the entire nanocrystal, indicating epitaxial growth (reproduced with permission from [828])

the type CdSe–CdS were obtained [830]. CdSe–ZnS and CdSe–ZnSe nanocrystals were prepared similarly [831].

Peng and coworkers [833] have adapted the successive ion layer adsorption and reaction (SILAR), an in-solution deposition method derived from the atomic layer epitaxy method used in growing thin films in vacuum, to prepare CdSe–CdS nanocrystals on gram scale batches using ordinary precursors such as CdO, S, and Se. Wei et al. [228] report a green synthetic route for CdSe/CdS core–shell nanoparticles in aqueous solution using Se powder as the selenium

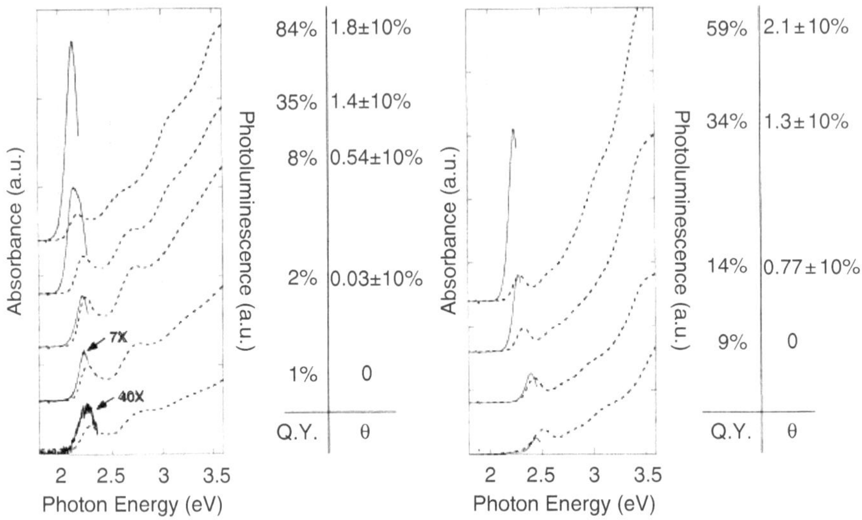

Fig. 5.2. Absorption (*dashed*) and photoluminescence (*solid*) spectra of two series of CdSe–CdS nanocrystals. Spectra were taken after successive injections of CdS stock solution. The increase in quantum yield and of coverage of CdS with each injection is also shown. Q.Y.: quantum yield of photoluminescence. θ: number of monolayers of shell growth. All spectra were taken at a concentration corresponding to an optical density (OD) of roughly 0.2 at the peak of the lowest energy feature in the absorption spectrum. 3.0 nm CdSe core diameter series (*left*); 2.3 nm CdSe core diameter series (*right*) (reproduced with permission from [828])

source. This route is of interest as it uses common reagents to make water soluble nanocrystals. Water dispersed CdTe–CdS nanocrystals with high PL (75% quantum yield) was prepared by the microwave technique [832].

Weller and coworkers [834] have prepared CdSe–CdS nanocrystals by reacting CdSe nanocrystals with H_2S. The core nanocrystals were prepared using different precursors such as dimethyl cadmium and cadmium acetate. The properties of the resulting core–shell nanocrystals were compared. This scheme has been extended to prepare nanocrystals of the form (CdSe–CdS)–ZnS and (CdSe–ZnSe)–ZnS [835]. The nanocrystals with a ZnS outer layer are highly luminescent. This method of growing two shell layers can be extended to grow shell material (on cores) with a higher degree of lattice mismatch than in simple core–shell nanocrystals. Cao and Banin [836, 837] have successfully coated InAs nanocrystals with shells of InP, GaAs, CdSe, ZnSe, and ZnS. Using the shell layers, the bandgap of InAs can be tuned in the near-IR region. Highly luminescent multishell nanocrystals of the composition CdSe-core CdS/$Zn_{0.5}Cd_{0.5}S$/ZnS-shell were prepared by successive ion layer adhesion and reaction technique [838]. The growth of the shell was carried out on monolayer at a time by alternating injections of cationic and anionic precursors into the reaction mixture with core nanocrystals.

There has been relatively less focus on the growth of a shell layer with a narrower bandgap. All the attempts have been carried out in an aqueous medium using either polyphosphate or PVP as capping agents. An early report by Fendler and coworkers [823] has explored the adaptation of the polyphosphate method to synthesize CdS and CdSe composite nanocrystals (see Fig. 5.3). Other systems studied include PVP capped CdS–PbS [839] and polyphosphate capped HgS–CdS [840]. A pronounced red shift of the absorption features and a slight enhancement of fluorescence intensity was observed. Water soluble CdSe–CdS core–shell nanocrystals with dendron carbohydrate anchoring groups were prepared. The nanocrystals retain 60% of the PL value of the original CdSe–CdS nanoparticles [841].

Despite the advances in synthetic schemes, several aspects of core–shell nanocrystals are not fully understood. The factors that determine epitaxial growth and defect migration to the shell layer, the nature and extent of the overlap of the electronic structures and its dependence on the bandgap mismatch, the role of lattice mismatch are some issues.

5.1.2 Metal–Metal

Metal on metal core–shell structures provide the means for generating metal nanocrystals with varied optical properties. The dielectric constant of the medium surrounding a metal nanocrystal can be varied in an extended region if one could cap a metal core with a metal shell instead of an organic layer. A noble metal layer on a transition metal core can lend stability from oxidation. In addition, the use of a seed as core layer could lower the reduction potential and permit easy reduction of the metal that forms the shell.

Morriss and Collins [842] prepared Au–Ag nanocrystals by reducing Au with P by the Faraday method and Ag with hyroxylamine hydrochloride. They observed a progressive blue shift of the plasmon band accompanied by a slight broadening. For sufficiently thick shells, the plasmon band resembled that of pure Ag particles. Large Au nanoparticles prepared by the citrate method was used as seeds for the reduction of Ag nanocrystals using ascorbic acid, with CTAB as the capping agent [843]. Au–Ag as well as Ag–Au nanocrystals were prepared by the sequential reduction using sodium citrate [844]. Mirkin and coworkers [845] have coated Ag nanocrystals with a thin Au shell to provide stability against precipitation under physiological conditions. A thin shell has little effect on the optical properties. Henglein and coworkers [846–853] have synthesized core–shell nanocrystals by radiolytic means. Thus, Au–Cd [846], Au–Tl [846], Au–Pb [846, 847], Au–Hg [848], Au–Sn [849], Ag–Pb [850, 851], Ag–Cd [852], Ag–In [850], and Ag–Hg [853] were obtained. In most cases, with increasing shell thickness, the plasmon band progressively acquired the characteristics of the shell. This change often involves a blue shift. In some cases, the metal in the shell alloys with the core metal producing anomalous shifts. Buhro and coworkers [854] used small Au nanocrystals capped with triphenyl phosphine as seeds and have deposited shell layers of Bi, Sn, and In,

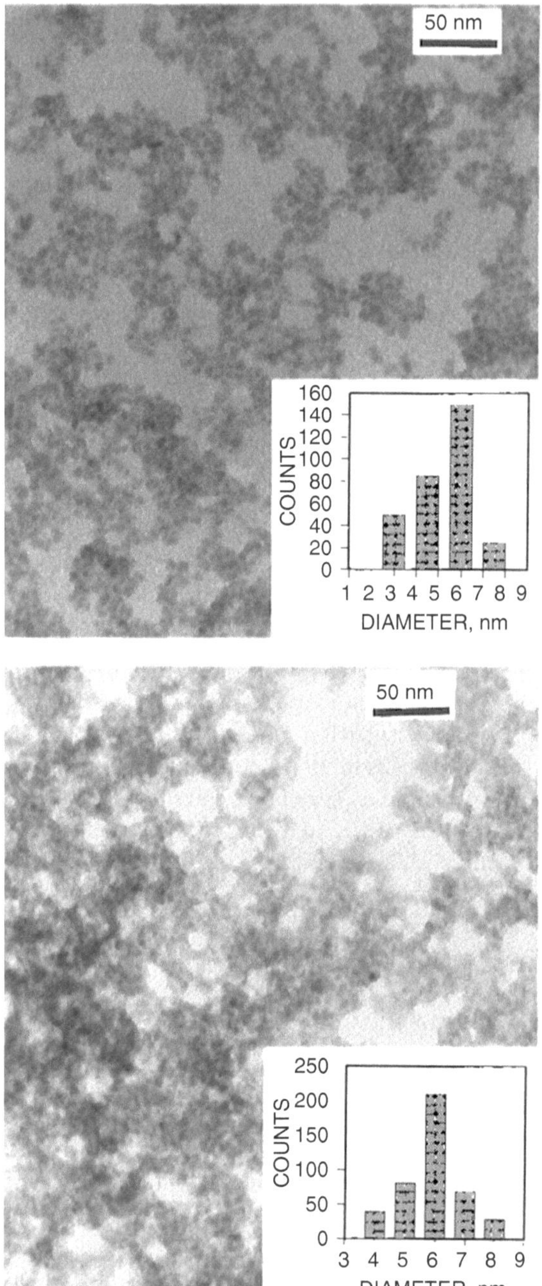

Fig. 5.3. TEM images and histograms of core–shell (**a**) CdS–CdSe and (**b**) CdSe–CdS nanoparticles (reproduced with permission from [823])

using $Bi[N(SiMe_3)_2]_3$, $Sn(NMe_2)_2$, and $In(C_5H_5)$. The synthesis, based on the method of Chaudret [301–303], was carried out at low temperatures to avoid heterogenous nucleation. Lambert and coworkers [855] have coated Au particles with a thin layer of Pd and found an unusual catalytic activity. EXAFS and other structural studies reveal that the Pd shells are structurally different from that of bulk Pd. The catalytic ability was attributed to electronic interaction between the core Au nanocrystals and shell Pd nanocrystals.

Reactive magnetic nanocrystals are rendered passive and made easy to handle by coating them with a layer of noble metals. A layer of Ag was grown in situ on Fe and Co nanocrystals synthesized using reverse micelles [856]. A similar procedure was used to coat Au as well [857]. Fe–Au nanocrystals are also prepared by sequential citrate reduction followed by magnetic separation [858]. Fe–Au nanoparticles were synthesized by a wet chemical procedure involving laser irradiation by Fe nanoparticles and Au powder in a liquid medium. The nanoparticles were superparamagnetic with a blocking temperature of 170 K [859]. A transmetallation reaction in which Pt displaces Co from the surface of Co nanocrystals was used to prepare Co–Pt nanocrystals. Sobal et al. found that the Ag core in Ag–Co nanocrystals protects Co from oxidation [860]. Differences in the electron density between Co and Ag have permitted direct imaging of the core–shell structure by TEM (see Fig. 5.4). Pt–Cu core–shell nanoparticles showing high activity for NO_x reduction were

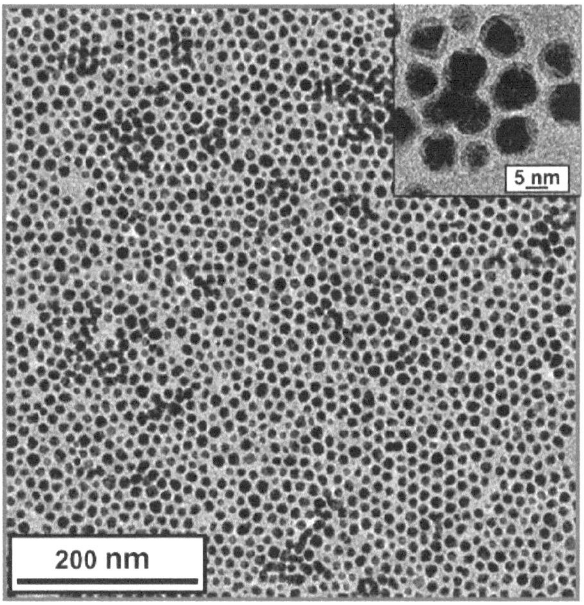

Fig. 5.4. TEM image of Ag–Co particles deposited on a carbon grid. The core–shell structure is visible at the higher magnification image shown in the inset (reproduced with permission from [860])

reported [861]. Magic nuclearity Pd_{561} nanocrystals were used as seeds to effect reduction of Ni in alcohol at lower temperatures. Pd_{561} nanocrystals uniformly coated with Ni were monodisperse with the magnetic moments scale varying with the Ni content [862, 863]. This method can be extended to prepare trilayered particles of the type (Pd–Ni)–Pd.

5.1.3 Metal–Oxide, Semiconductor–Oxide, and Oxide–Oxide

A dielectric oxide layer such as silica is useful as shell material because of the stability it lends to the core and its optical transparency. The thickness and porosity of the shell are readily controlled. A dense shell also permits encapsulation of toxic luminescent semiconductor nanoparticles. The classic methods of Stöber and Iler for solution deposition of silica are adaptable for coating of nanocrystals with silica shells [864, 865]. These methods rely on the pH and the concentration of the solution to control the rate of deposition. The natural affinity of silica to oxidic layers has been exploited to obtain silica coating on a family of iron oxide nanoparticles including hematite and magnetite [866–870]. The procedures are mostly adaptations of the Stöber process. Oxide particles such as boehmite can also be coated with silica [871]. Such a deposition process is not readily extendable to grow shell layers on metals. The most successful method for silica encapsulation of metal nanoparticles is that due to Mulvaney and coworkers [872–875]. In this method, the surface of the nanoparticles is functionalized with aminopropyltrimethylsilane, a bifunctional molecule with a pendant silane group which is available for condensation of silica. The next step involves the slow deposition of silica in water followed by the fast deposition of silica in ethanol. Changes in the optical properties of metal nanoparticles with silica shells of different thicknesses were studied systematically [873–875]. This procedure was also extended to coat CdS and other luminescent semiconductor nanocrystals [542, 876–879].

Silica-coated nanocrystals are more resistant to oxygen and photodegradation. Ag/Au–SiO_2 particles with dye-labeled SiO_2 shell exhibit fluorescence enhancement [880]. The complexity of the two-step process to coat nanocrystals has fueled a search for alternate single step processes. Successive reactions in inverted micelles are used to prepare Ag–SiO_2 [881]. Liz-Marzan and coworkers [882] carried out the reduction of Ag ions with DMF in the presence of aminosilane to obtain Ag–SiO_2 nanocrystals. Pradeep and coworkers [883,884] coated Au and Ag nanoparticles with TiO_2 and ZrO_2 in a singe-step process. The oxide-coated nanocrystals were stable under extreme conditions. Dumbbell shaped Au–Fe_2O_3 nanoparticles were prepared by the decomposition of $Fe(CO)_5$ on the surface of Au nanoparticles, followed by oxidation [885]. Core–shell nanoparticles with a ferrimagnetic $CoFe_2O_4$ core and an antiferromagnetic MnO shell were obtained by a high-temperature decomposition route with seed mediated growth [886].

The charge doping scheme of Guyot–Sionnest has been extended to CdSe–ZnS nanocrystals [529, 530]. It is suggested that by proper tuning of the core

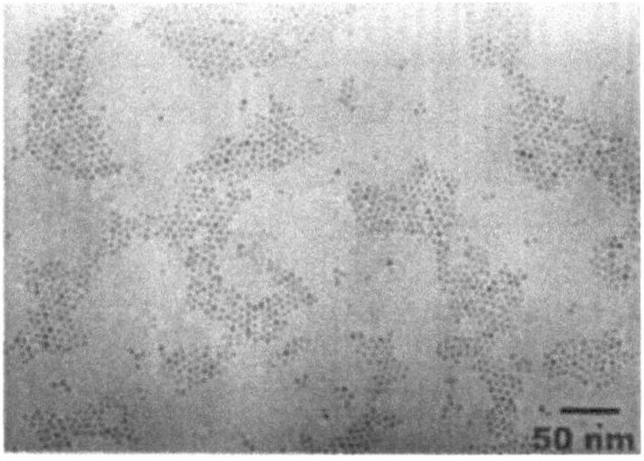

Fig. 5.5. TEM images of the two-dimensional arrays formed by octanethiol-coated (Pd_{561}–Ni_{3000})–Pd_{1500} nanocrystals (reproduced with permission from [862])

and shell layer, a nanocrystal whose fluorescence can be completely quenched by adding one electron can be obtained. Nanocrystals exhibiting such behavior could be used to make functional electrochromic devices.

5.2 Assemblies of Core–Shell Nanocrystals

Ordered assemblies of core–shell nanocrystals in two dimensions are not well-known. However, it is expected that core shell nanocrystals, especially metal–metal and semiconductor–semiconductor types can be obtained following the same methods used to assemble nanocrystals. For example, PVP capped Pd_{561}–Ni_x (x up to 10,000) were derivatized with alkanethiols and alkylamines, and were found to self-assemble into two-dimensional lattices [862]. Double shell nanocrystals of the form $Pd_{561}Ni_{3000}Pd_{1500}$ were organized into two-dimensional arrays (see Fig. 5.5) using alkane thiols as capping agents [862]. Two-dimensional arrays of CdSe–ZnS nanocrystals were obtained using chaperonin as the template [610].

6

Applications

6.1 Introduction

Applications of nanocrystals are being increasingly realized with vigorous efforts aimed at generating nanocrystal-based commercial products. Many proof of concept experiments are also underway to assess if nanocrystals satisfy some of the predictions related to applications. While major products based on nanoelectronics may not become commercial in the near future, some of the applications related to sensors, diagnostics and particulate technology are already being exploited. Typical particulate technologies are in areas of catalysis, textiles and cosmetics. In the following sections, we examine the applications both real and potential, in areas such as electronics, optical devices, computing, and biology.

6.2 Nanocrystals as Fluorescent Tags

Quantum dots possess definite advantages over conventional fluorescent dyes such as rhodamine [543, 877, 887–889]. They are highly photostable, possess large stoke shifts and emit in a narrower range (with emission peak widths about a third of the width seen in the case of molecular species). In Fig. 6.1 the spectral properties of rhodamine are compared with those of ZnS–CdSe nanoparticles to highlight how multiple narrow, quantum dot emissions can be used in the same spectral window as that of an organic or genetically encoded dye. It is possible to simultaneously obtain intense emission at several different wavelengths by exciting at a single wavelength by the use of nanocrystals of the same material but with different sizes. Furthermore, emission from quantum dots is not susceptible to quenching.

Quantum dots capable of emitting light have been covalently attached to biomolecules by functionalizing their surface with carboxyl or related groups. For example, transferrin and IgG have been attached to CdSe–ZnS nanocrystals functionalized with mercaptoacetic acid [543]. Figure 6.2 illustrates the utility of quantum dots of various semiconductors as fluorescent tags. Most biological usage falls in the visible–near infrared region [887]. It is also possible

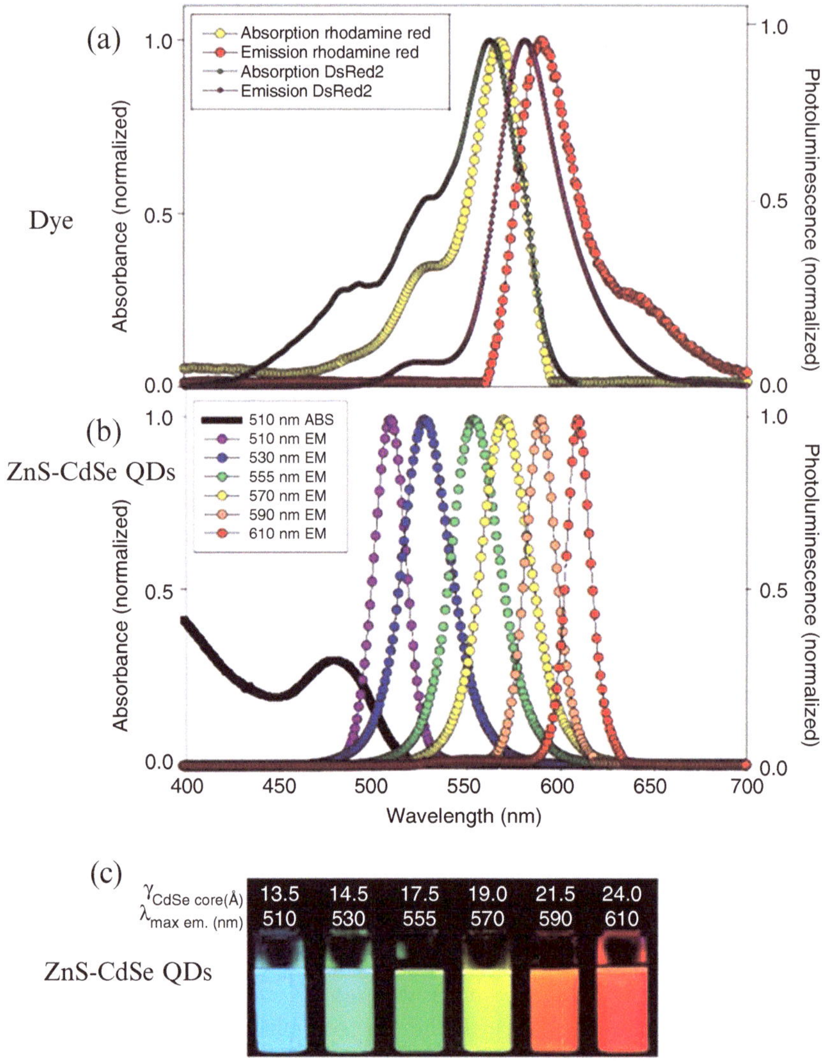

Fig. 6.1. Absorption and emission spectra of (**a**) rhodamine red and a genetically encoded protein (DsRed2), (**b**) six different ZnS–CdSe quantum dot dispersions, (**c**) A photograph demonstrating the size-tunable fluorescence properties of quantum dot dispersions (reproduced with permission from [887])

to attach quantum dots to biomolecules electrostatically. More sophisticated and specific attachment could be obtained by covalently linking a designer peptide incorporating a receptor sequence or moiety. In this context, nanomaterials such as heterodimers of CdS and FePt nanocrystals (see Fig. 6.3) exhibiting both superparamagnetism and fluorescence properties are potentially attractive in bioapplications [890].

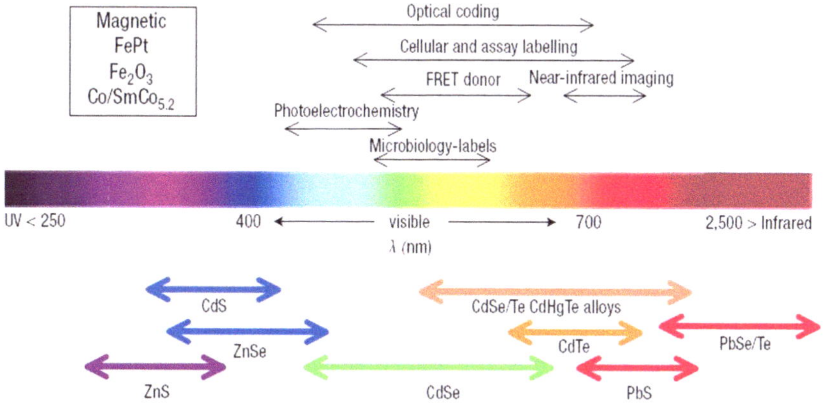

Fig. 6.2. The emission wavelength regions of typical semiconductor quantum dots and their areas of biological interest. Representative materials used for creating magnetic quantum dots are shown in the inset (reproduced with permission from [887])

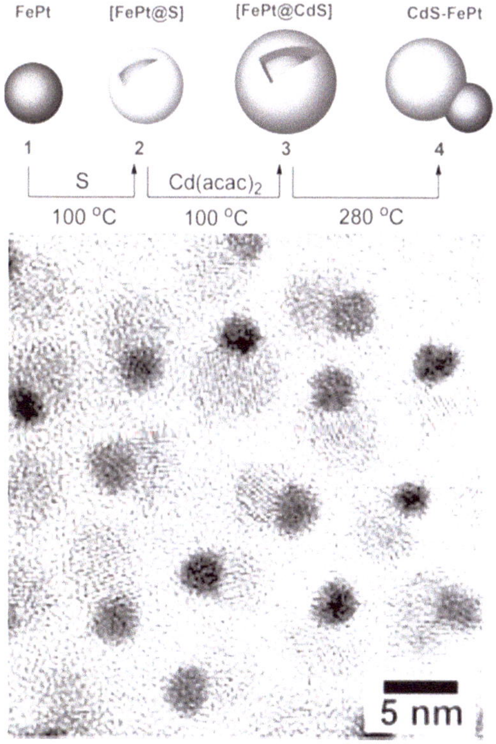

Fig. 6.3. Heterodimers of CdS and FePt nanocrystals: FePt–CdS core–shell nanoparticle turning into a heterodimer upon annealing is shown in the scheme. TEM image shows crystalline CdS (zinc blende structure) anchoring to disordered FePt nanocrystal (reproduced with permission from [890])

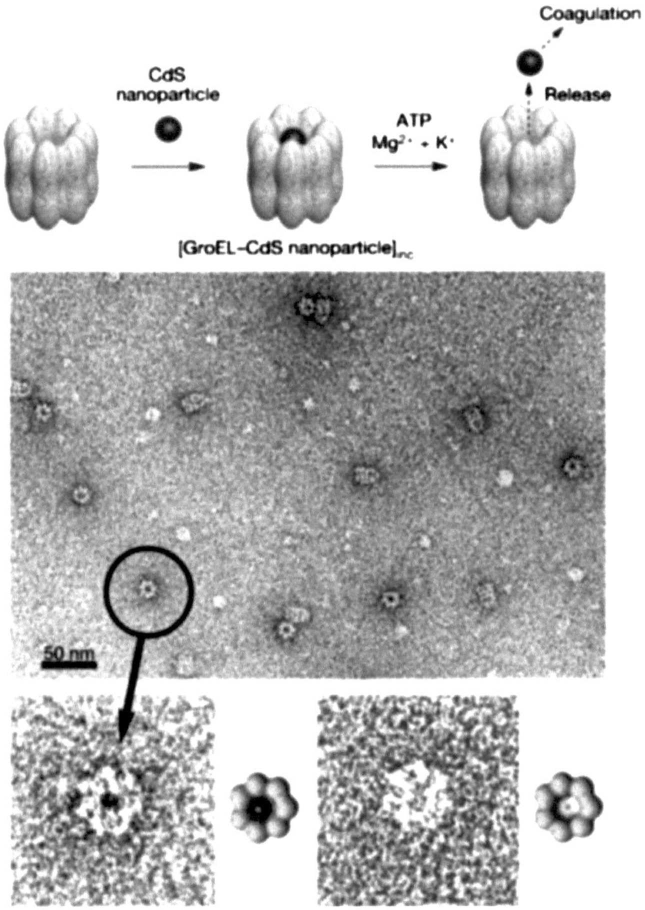

Fig. 6.4. GroEL, a cylindrical chaperonin protein complex that functions in refolding denatured proteins in vivo, complexes with a CdS nanocrystal holding in the cavity and releasing it when triggered by ATP. TEM image showing GroEL–CdS complex before and after the release (reproduced with permission from [892])

Several in vivo [544,545,891,892] and in vitro [893–895] studies have been carried out using quantum dots as either specific or nonspecific labels. In an elegant work due to Ishii et al. [892], CdS quantum dots bound in the central pocket of a protein complex were shown to get released on the addition of ATP (see Fig. 6.4). It appears that quantum dots are likely to replace conventional dyes in the near future. The enhanced photostability of quantum dots also makes it possible to follow binding events in real time. Furthermore, the usage of single-wavelength excitation to obtain emission at different wavelengths has applications in combinatorial searches. QD labeling permits extended visualization of cells under continuous illumination as well as multicolor imaging (see Fig. 6.5).

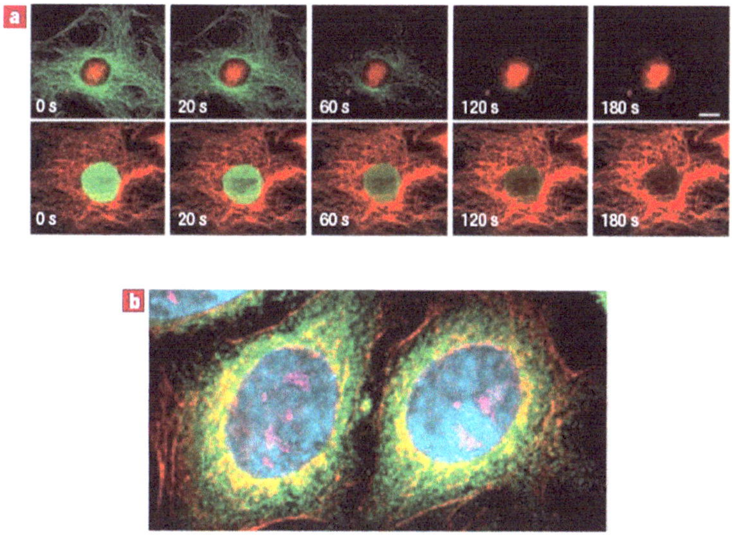

Fig. 6.5. Quantum dot resistance to photobleaching and multicolor labeling. (**a**) *Top row*: nuclear antigens labeled with quantum dot 630-streptavidin (*red*), and microtubules labeled with AlexaFluor 488 (*green*) simultaneously in a 3T3 cell. *Bottom row*: microtubules labeled with quantum dot 630-streptavidin (*red*), and nuclear antigens stained green with Alexa 488. Continuous exposure times in seconds are indicated. The resistance of quantum dots to photobleaching under continuous illumination is apparent. (**b**) Pseudocolored image depicting quantum dots of five-colors staining fixed human epithelial cells. Cyan corresponds to quantum dots labeling the nucleus, magenta quantum dots labeling Ki-67 protein, orange quantum dots labeling mitochondria, green quantum dots labeling microtubules and red quantum dots labeling actin filaments (reproduced with permission from [887])

6.3 Nanocrystal-Based Optical Detection and Related Devices

The optical response of metal nanocrystals to changes in the dielectric constant of the surrounding medium has been used to design probes that respond to a specific event such as oligonucleotide pairing with its complimentary sequence, antibody–antigen binding or in general, protein–protein interactions [77, 896, 897]. A pioneering study by Mirkin and coworkers has initiated this area of research [77]. Gold nanocrystals of 13 nm diameter were made water soluble by derivatizing the surface with a thiol terminated oligonucleotide sequence. A colorimetric response (color change from red to blue) was obtained when a complimentary DNA strand was added. This color change can visibly indicate base-pair mismatches. The process is reversible, and the blue-colored nanocrystalline dispersion regains its original red color when the chains are sliced apart by heating. It is possible to observe a color change by spotting the nanocrystals on silica gels (see Fig. 6.6) [898]. DNA fragments melt (lose their

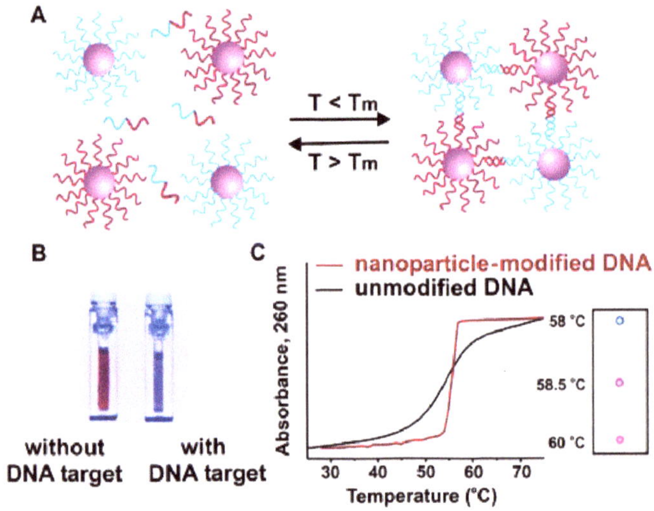

Fig. 6.6. In the presence of complimentary target DNA, oligonucleotide functionalized Au nanocrystals aggregate (**a**), resulting in a change of solution color from red to blue (**b**). The aggregation process can be monitored using UV–visible spectroscopy or by spotting on silica gel (**c**) (reproduced with permission from [929])

rigid conformation) over a much narrower temperature range when bound to metal particles than to conventional dyes [899]. Different color change patterns are obtained by using Au–Ag alloy or Ag–Au core–shell nanocrystals. This scheme of detection has received much attention. Small Au nanoparticles are shown to exhibit high photoluminescence upon irradiation with femtosecond pulses of 790 nm light suggesting that the metal nanoparticles are good alternatives to fluorophores or semiconductor nanoparticles for biological labeling and imaging [900].

Ag nanotriangles generated by nanosphere lithography have been used to detect biotin–streptavidin binding by immobilizing biotin on the particles [901, 902]. Treatment with streptavidin solutions with concentrations in picomolar range was sufficient to trigger the response. Nie and coworkers [903] have suggested a novel replacement for molecular beacons used to signal binding events through fluorescence quenching. In their scheme, either ends of a oligonucleotide is functionalized – one end with a thiol group and the other end with a fluorophore. The thiol group binds to the Au nanocrystal by covalent linkage while the fluorophore loosely attaches itself to the surface of the nanocrystal; thereby losing its fluorescence. The oligonucleotide adapts a hairpin like conformation. When a complimentary DNA sequence is added, the fluorophore is detached from the surface regaining its fluorescence and thus acting as a beacon. Inhibition assays based on fluorescence resonance energy transfer (FRET) between streptavidin-conjugated quantum dots and biotin-related Au nanoparticles has been described [904]. It has been suggested that larger nanoparticles could be very sensitively detected by measuring the light scattered by them [905–907]. The intensity of the scattered light by a single 80 nm Au particle has been compared to the total light emitted by 10^6 fluoresceing molecules. Yguerabide and coworkers [905–907] have pioneered the use of large particles with diameters of ∼60 nm for carrying out tracer type biochemical assays. Lippitz et al. have shown that third harmonic signals can be generated from gold colloids with diameters of 40 nm, showing a potential use for tracking of single-biomolecules [908].

Trace analysis of molecules in biochemistry and medicine has been realized using surface enhanced Raman scattering (SERS) with colloidal particles, especially of Ag and Au. When adsorbed on particle surfaces, molecules exhibit an enhanced Raman signal usually of several orders of magnitude higher compared to free molecules [909]. Nanoparticle over smooth electrode (NOSE) has emerged as a method for SERS active substrates, where a nanoparticle system carrying the analyte is deposited over a designated surface by drop coating to enhance plasmon coupling [910].

In addition to responding to the changes in the surrounding medium, absorption spectra contain other important information as well. Dipolar coupling interactions in metal nanocrystal assemblies such as linear rows, lead to strong optical anisotropy. The transmitted light intensity thus depends on whether it is polarized parallel or perpendicular to the rows of nanocrystals. Such material could be useful as polarizing filters. Dirix et al. [911] have pioneered a simple method of preparing filters by stretching a polyethylene film impregnated with Ag nanocrystals. Another example is planar assembly of CdSe nanocrystals emitting linearly polarized light in the plane of the assembly [912]. Atwater and coworkers [913] have suggested that a coupled plasmon mode could lead to coherent propagation of electromagnetic energy, i.e., a row of metal nanocrystals acting as a nanoscale waveguide. Using an array generated by means of e-beam lithography, they have provided a proof of concept demonstration. Nanoparticle assemblies connected by polymers can act as molecular springs and nanothermometers. Surface plasmon resonance and exciton–plasmon interaction are responsible for the nanothermometer function [914].

6.4 Biomedical Applications of Oxide Nanoparticles

Magnetic oxide nanoparticles such as iron oxide are considered to be biocompatible as they possess no known toxicity [915]. Superparamagnetic oxide nanoparticles have therefore found several biomedical applications. By tailoring the ligand shell, the magnetic particles can be attached to a specific target molecule in a solution of different entities. The target can then be separated by the use of a magnetic field. Such aids are important in biochemical experiments where very low concentrations are generally employed. For example, by employing red blood cells labeled with iron oxide nanoparticles, the sensitivity of detection of malarial parasite is shown to be enhanced [916].

It has been proposed that hyperthermia – magnetic field induced heating of superparamagnetic particles – could be used to destroy diseased cells and thereby treat cancer. The particles are dispersed in the affected cells and an external AC magnetic field is applied to heat the particles and destroy the cells. Numerous studies have been carried out toward achieving this objective [917, 918]. A magnetic nanoprobe consisting of superparamagnetic nanoparticles coated with a specific molecule of interest has been used to study molecular interactions in live cells [919]. Magnetic nanoparticles also find applications as contrast enhancing agents in Magnetic Resonance Imaging. This technology has been commercialized and a nanoparticle-based contrast agent called Feridex I.V is commercially available [915].

6.5 Optical and Electro-Optical Devices

Devices based on composites of nanocrystals with polymers or polyelectrolytes have been fabricated to exploit electroluminescent and related properties of nanocrystalline media. These devices utilize thin films of composites obtained by layer-by-layer deposition using self-assembly, drop-casting, or spin-casting methods.

Sizable photocurrents are obtained from CdS nanocrystals anchored to Au substrates using dithiols [618]. Solar cells with efficiency comparable to the commercial cells have been made using poly(2-hexylthiophene)–CdSe nanorod multilayers [920]. Organic solar cells have been fabricated using porphyrins and fullerene units along with Au nanoparticles deposited on nanostructured SnO_2 electrodes [921]. CdSe nanocrystals linked to mesoporous TiO_2 films exhibit injection of electrons from the nanocrystals to the TiO_2 films upon visible light excitation. Photocurrent is generated by collecting the injected electrons at a conducting electrode [922]. HgTe nanocrystals increase photon-harvesting efficiency of hybrid solar cells in the 350–1,500 nm regions. Devices fabrication show photon to current efficiencies up to 100% at 550 nm [923]. Photocurrent generation as well as electroluminescence have been found in devices made up of CdS nanocrystals and polymers such as poly(3-hexylthiophene) and poly[2-methoxy-5-2(2-ethylhexoxy)]-1,4-phenylene vinylene [924]. Electroluminescence has been seen in devices consisting of CdSe [925] and CdSe–ZnS [926] nanocrystals embedded in films of poly(vinylcarbazole) and a oxidiazole derivative. In an interesting study, Gudiksen et al. [927] observed electroluminescence from a single CdSe nanocrystal transistor. The light emission occurred when the bias voltage exceeds the band gap of CdSe. Light emitting diodes have been made with CdSe [928] and CdSe–CdS [929] nanocrystals (see Fig. 6.7). Full color emission has achieved by using composites of CdSe–ZnS and CdS–ZnS in polylaurlymethacrylate [930]. The polymerization reaction was carried out in the presence of the nanocrystals to achieve optimal incorporation of the particles in the film. The right configuration to yield full color emission

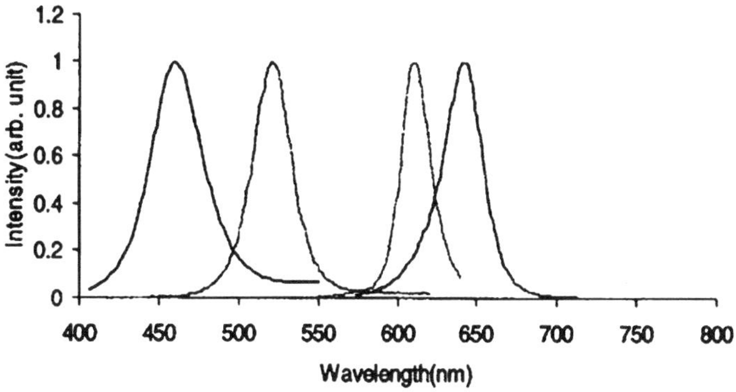

Fig. 6.7. Emission spectra of semiconductor nanocrystal polymer composites that were used to make a white light emitting film. (*From left*) CdS–ZnS core–shell nanocrystals (diameter 3 nm), CdSe–ZnS of diameters 1.3, 2.3, 2.8 nm, respectively. The excitation was from a single source (mercury lamp with emission wavelength of 365 nm) (reproduced with permission from [929])

has been attempted [931]. In this study, CdTe multilayers were obtained on ITO glass using PDDA as the crosslinker. By varying the diameter of the nanocrystals in different layers, electroluminescence of different colors could be obtained. Efficient photodetectors based on poly[2-methoxy-5-(2-ethylhexyloxy)-1,4-phenylene vinylene] and PbSe nanocrystal composites have been fabricated [932]. The observed photocurrent gain is attributed to carrier multiplication in PbSe nanocrystals via multiple exciton generation and efficient charge transport through the polymer matrix.

Lasers have been fabricated by using composites containing titania and CdSe, CdSe–ZnS, or CdS–ZnS core–shell nanocrystals with strong confinement [933, 934]. By varying the size of the CdSe nanocrystals from 1.7 to 2.7 nm or by using CdSe–ZnS nanocrystals, the wavelength of stimulated emission is varied. The novelty is that lasers of different colors can be obtained using a single method of fabrication.

6.6 Dip-Pen Nanolithography with Nanocrystals

Dip-pen nanolithography (DPN) is a atomic force microscopy-based lithographic technique that can be used to generate patterns with dimensions extending from a few nanometers to micrometers. In this technique, a water meniscus, deliberately created by slowly scanning an atomic force cantilever on a substrate, is used to deliver molecules or other material previously coated on the tip to the surface. The cantilever acts as a "nib" delivering previously coated "ink" to the substrate. The "ink" usually binds chemically to the substrate. The process is schematically illustrated in Fig. 6.8 for the case of Au nanocrystals. Different DPN-based methods have been used to generate

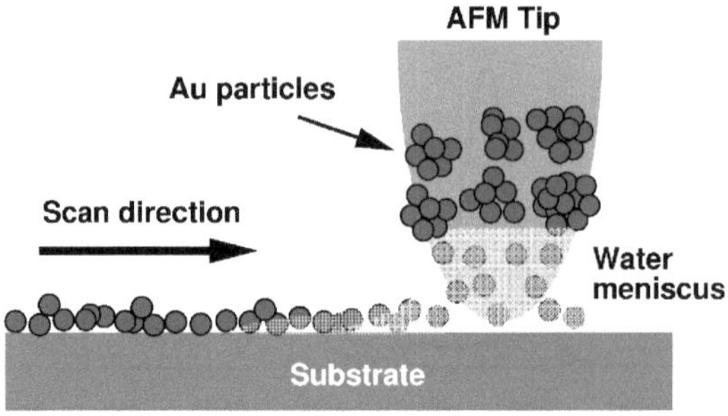

Fig. 6.8. Schematic illustration of the "dip-pen process." Nanocrystals deposited on the cantilever dissolve in the water meniscus between the tip and the substrate. The dissolved nanocrystals are transferred to the substrate, when the tip is slowly scanned across the surface. Aggregates of nanocrystals are formed on the substrate upon evaporation of the water layer

nanoscopic patterns of nanocrystals. By using bifunctional thiols as inks, colloids, proteins, and other macromolecules can be tethered to the surface at specific regions [449, 938–940]. A DPN based procedure has been used to deposit small volumes of an organosol containing Au nanocrystals leading to the formation of circular nanocrystal patterns on the substrate [941]. Liu and coworkers [940] have suggested two methods to obtain linear patterns of Au nanocrystals on Au substrates. The first method involves creating trenches in a monolayer of thiols on an Au substrate in a liquid medium containing Au nanocrystals coated with a mixture of mono- and di-thiol molecules. The second method involves direct deposition from an AFM tip, coated with nanocrystals, in trenches in thiol monolayers, created by the same tip at high contact forces. Hydrosol inks of THPC coated Au nanocrystals, PVP covered Pd nanocrystals and citrate stabilized magnetic γ-Fe_2O_3 nanocrystals have been used to generate patterns of nanocrystals on mica and silicon substrates (see Fig. 6.9) [942,943]. In the case of the iron oxide nanocrystals, the patterns were further studied by means of magnetic force microscopy and nanospectroscopic electron imaging techniques (see Fig. 6.10). Recently, DPN has been effectively combined with conventional micro and nanofabrication techniques to obtain a macroscopic electronic circuit that addresses a single Au nanocrystal [944]. DPN has been used to add local chemical functionality (in the form of specific DNA sequences) to the edges of lithographically defined electrodes a few nanometers apart. The procedure is schematically illustrated in Fig. 6.11. Clearly, DPN can be used for generating functional patterns of nanocrystals and is set to emerge as a powerful method for generating nanoscopic patterns of nanocrystals.

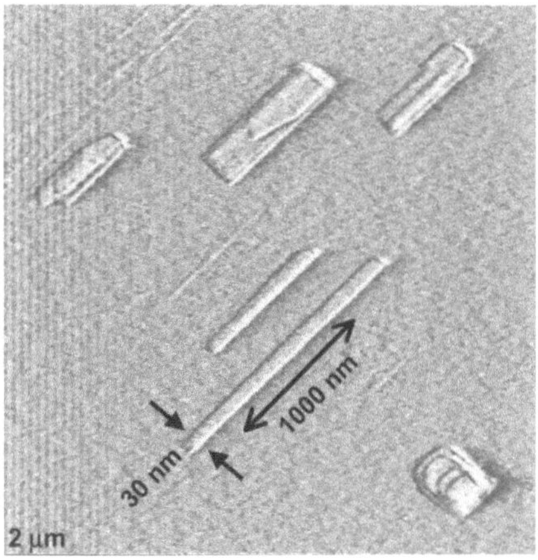

Fig. 6.9. AFM images showing various patterns made of Au nanocrystals. The patterns were drawn by scanning the cantilever over the surface with a velocity of 1 $\mu m\,s^{-1}$, under ambient conditions. Subsequent imaging was carried out over larger areas at higher velocity (reproduced with permission from [943])

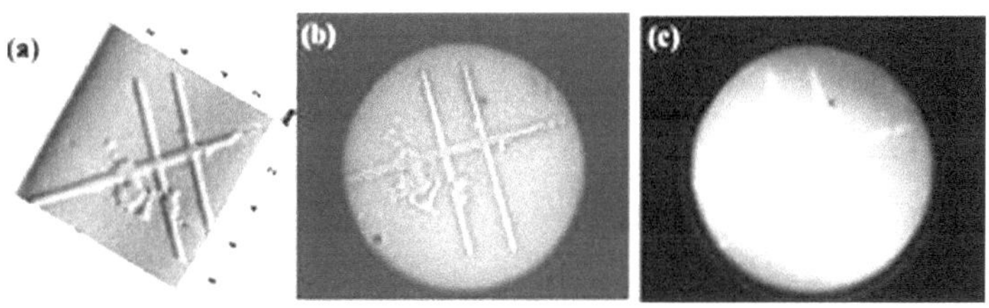

Fig. 6.10. (**a**) Tapping mode topographic atomic force microscopic image of a magnetic γ-Fe_2O_3 nanocrystal pattern along with (**b**) the corresponding mirror electron microscopy (MEM) image collected with electron beam of energy 0.8 eV and (**c**) X-ray photoemission electron microscopy image obtained by collecting the secondary electrons (energy 0.6 eV) following excitation with 712 eV (Fe L3 edge). The field of view in both cases is 10 μm (reproduced with permission from [942])

6.7 Nanoelectronics and Nanoscalar Electronic Devices

The single electron devices such as supersensitive electrometers and memory elements could be fabricated using nanocrystals that has been one of the expectations of some practitioners in the field. This is in line with the belief that nanoobjects could bring a new era in electronics, aptly named nanoelectronics. Proof of concept experiments to test chemically prepared nanocrystals in single electron devices have been carried out. For example, Murray and coworkers [945] have found that a single redox reaction taking place at the surface of Au nanocrystals induces a eightfold increase in its capacitance. The concept of single electron transistors (SET), where the change in state is brought about by charging the transistor with the smallest quanta of charge,

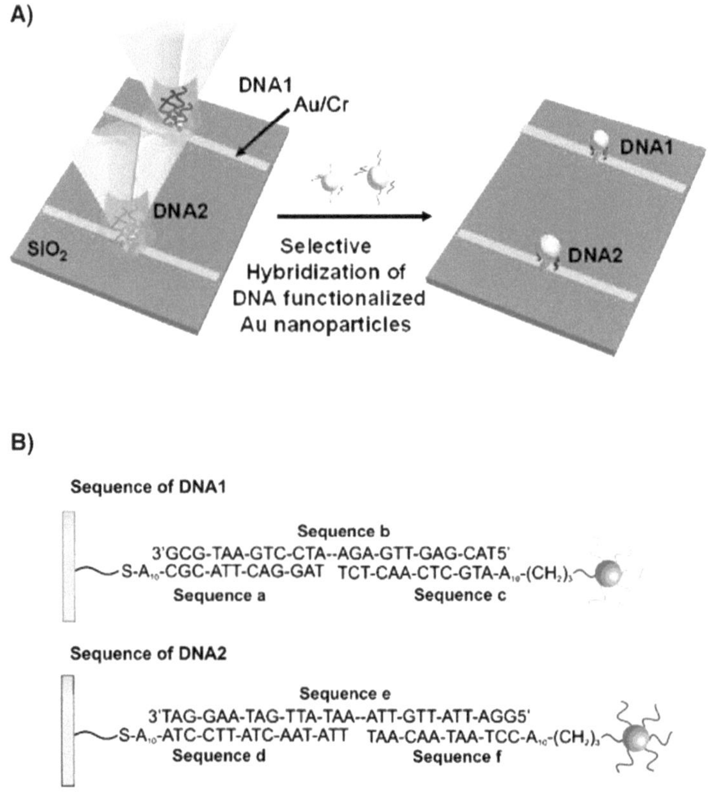

Fig. 6.11. (**a**) DNA-directed assembly of Au nanoparticles modified with oligonucleotides at the gap between metallic electrodes selectively patterned with DNA via DPN (**b**) schematic representation of two different DNA systems used to fabricate the devices (reproduced with permission from [944])

would be energy efficient transistors. Such transistors have been fabricated based on lithographically defined quantum dots. The computers of the future are to be made of SET. The only problem with implementing such a scheme with the current state of advancement is that it would require the computer to be operated at cryogenic temperatures. A nanocrystal with its high charging energy could be a key ingredient in making a room temperature SET. Indeed, SET have been fabricated with one or a few chemically prepared nanocrystals held between the electrodes (see Fig. 6.12) [946, 947].

Some of them are operable at room temperature [947]. It is hoped that such transistors can be built by a self-assembly process in the future.

Schiffrin and coworkers [948] obtained a nano-switch based on a layer of Au nanoparticles on a viologen moiety anchored to a gold substrate. The I–V characteristics of the Au nanoparticles obtained by in situ scanning tunneling spectroscopy (STS) revealed a dependence on the redox state of the viologen

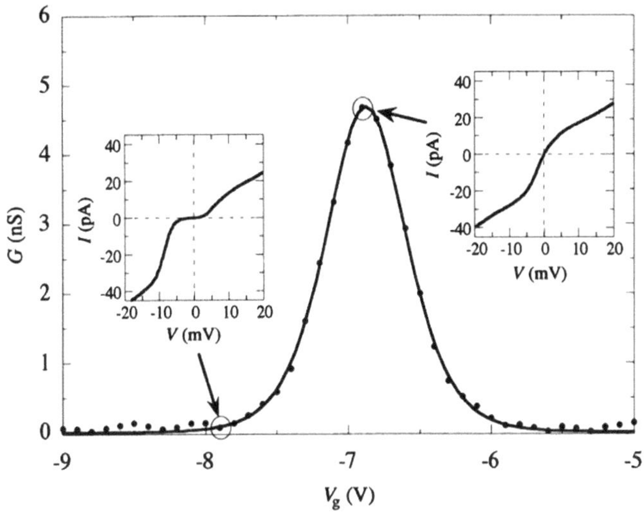

Fig. 6.12. Conductance, G, plotted against gate voltage, V_g, for a single nanocrystal transistor measured at $T = 4.2\,\mathrm{K}$. The conductance shows a peak when the charge state of the nanocrystal changes by one electron. The *dots* are the measured values; the *solid curve* is a fit to the data by using the standard Coulomb blockade model with a temperature $T = 5\,\mathrm{K}$. Insets show the I–V characteristics measured at the gate voltages indicated (reproduced with permission from [946])

underneath the nanoparticle. By electrochemically altering the redox state, the conductivity of the circuit could be made high or low. A STS study on two Pd nanocrystals linked with a conjugated thiol has been shown to act as a switch. The I–V spectra exhibit negative differential resistance indicative of switching behavior [949]. Besides such nanoelectronic appliances, a nanomotor consisting of a single metal nanocrystal sandwiched between mechanical arms has also been fabricated [950].

A more tangible application of nanocrystals is their use as vapor sensors. There have been a few attempts to obtain measurable electrical response for vapors of molecules that can adsorb to the surface of the nanocrystals. The molecules include thiols or simple solvent molecules such as toluene and ethanol. The sensor elements in these cases have been prepared by multilayer deposition [951] or spin-coating techniques [952–954]. Murray and coworkers [955] have used Cu ions and thiolacid protected nanocrystals to build such a sensor element (see Fig. 6.13). These devices produce a reversible and rapid response to different kinds of vapors (see Fig. 6.14). In other cases, sensitivity of parts per million concentration has been achieved [956]. For example, nanoparticles of ZnO are found to be excellent for sensing ethanol and hydrogen. Similarly, In_2O_3 nanoparticles are good sensors for Nitrogen Oxides. Luminescent semiconducting nanocrystals of materials such as CdTe may be useful for optical oxygen sensing [957].

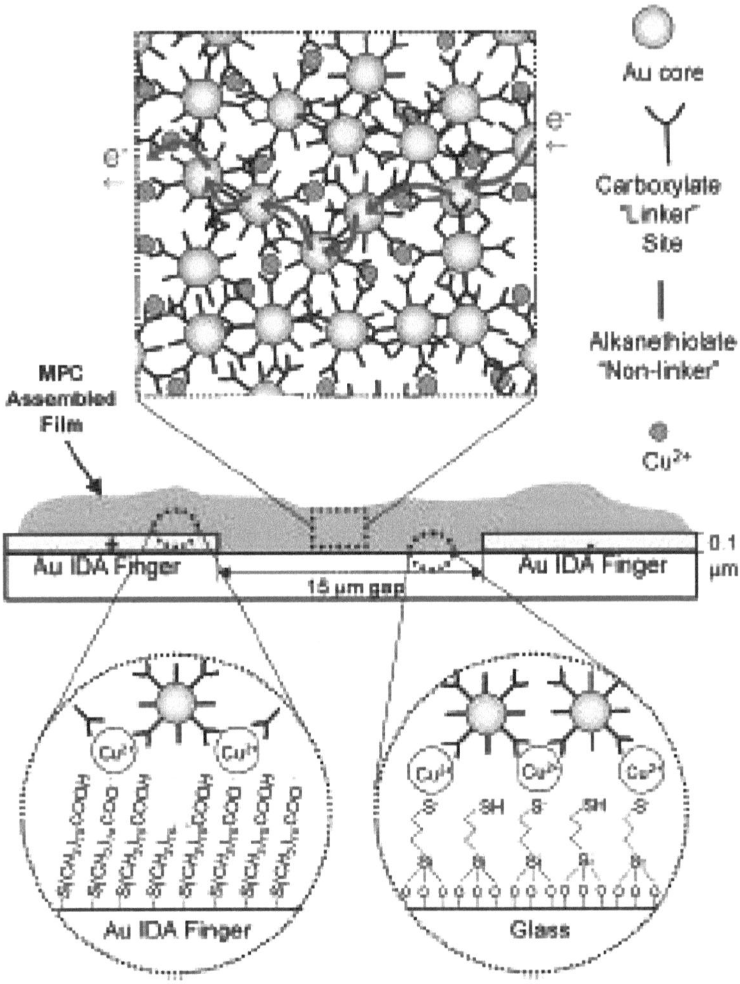

Fig. 6.13. Schematic illustration of a sensor element used to obtain the electrical response (reproduced with permission from [955])

Magnetic nanocrystals have been envisaged for use in recording. For example, ferromagnetic lattices made of FePt nanocrystals sustain high-density magnetization transition (see Fig. 6.15). An electrical transition induced by high electrical fields is observed in a device consisting of 2-naphtolenethiol-capped Au nanoparticle/polystyrene composite sandwiched between two Al electrodes, an observation with potential application in memory devices [958]. Memory effects have also been observed in polyaniline nanofiber/Au nanoparticle composites [959] and in CdSe nanocrystal arrays [960].

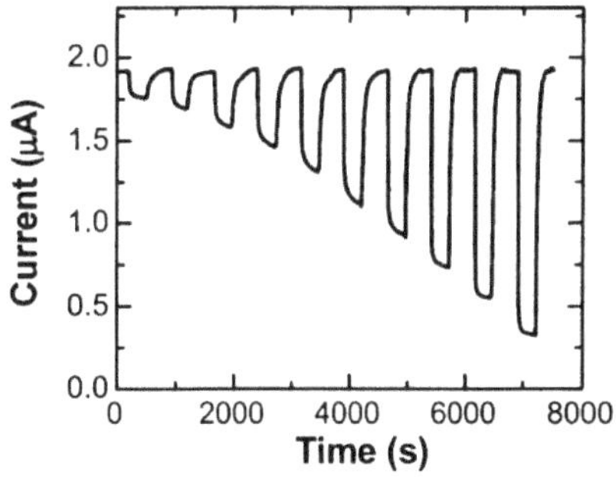

Fig. 6.14. Changes in the current over time when films were alternately exposed to nitrogen followed by ethanol vapor of increasing partial pressures. Ethanol vapor partial pressures were increased in fractions of 0.1 from 0.1 to 1.0 (reproduced with permission from [955])

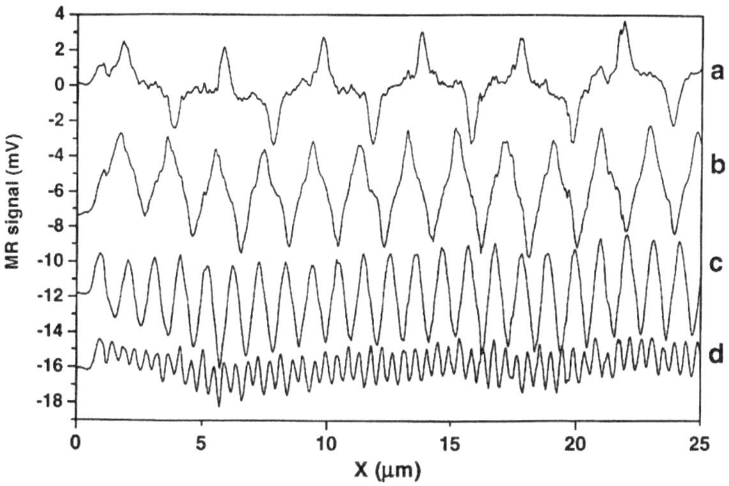

Fig. 6.15. Magneto-resistive (MR) read-back signals from written bit transitions in a array of 4 nm diameter $Fe_{48}Pt_{52}$ nanocrystals. The line scans reveal magnetization reversal transitions at linear densities of (**a**) 500, (**b**) 1040, (**c**) 2140, and (**d**) 5,000 flux changes per mm (reproduced with permission from [297])

6.8 Nanocomputing

Ordered arrays of nanocrystals can be thought of as arrays of SETs, where the electrostatic interaction between neighboring SETs acts as wireless communication means. It has been suggested by Korotkov [64] and Lent [961] that simple logical operations can be performed on a circuitry consisting of arrays of SETs in the form of chains or cells with suitable insulating spacers. An electric field applied in one direction polarizes the strings into either the 0 or the 1 state. Lent's scheme, named quantum cellular automata, instead uses a square cell consisting of five nanocrystals to denote the state of polarization. Snider et al. [962] reviewed the proposed schemes in the light of experimental efforts. Preliminary experiments are currently being pursued to evaluate such schemes. The realization that the fabrication ought to be based on self-assembly and that it is not capable of producing defect-free structures, has fueled a search for algorithms that can compute even with defective circuitry. Heath and coworkers [963] have developed Teramac, a defect-tolerant computer that works despite a high concentration of defects in its bank of microprocessors. A more radical solution called amorphous computing aims to "engineer prespecified, coherent behavior from cooperation of large numbers of unreliable parts interconnected in unknown, irregular and time varying ways" [964–966]. In this context, Hatano et al. [967] have examined the dependence of the tunnel coupling on the electron-spin and orbital components in neighboring quantum dots. This topic has been reviewed by van der Wiel et al. [968]. Special schemes have been proposed for quantum computation using the spin states of coupled single-electron quantum dots [969]. Cerletti et al. [970] have discussed practical and theoretical aspects related to quantum dot-based computing.

References

1. J. Jortner, C.N.R. Rao: Pure Appl. Chem. **74** 1491 (2002)
2. D.J. Barber, I.C. Freestone: Archaeometry **32** 33 (1990)
3. José-Yacamán, L. Rendón, J. Arenas et al.: Science **273** 223 (1996)
4. M. Faraday: Philos. Trans. R. Soc. London **147** 145 (1857)
5. A. Einstein, Ann. Phys. **17** 549 (1905)
6. G. Mie: Ann. Phys. **25** 377 (1908)
7. R. Gans: Ann. Phys. **31** 881 (1911)
8. R. Gans: Ann. Phys. **47** 270 (1915)
9. W. Ostwald: *Die Welt der Vernachlässigten Dimensionen* (Steinkopf, Dresden 1915)
10. Bredig: Z. Angew. Chem. **11** 951 (1898)
11. Donau: Monatsh **25** 525 (1905)
12. Zsigmondy: Z. Phys. Chem. **56** 65 (1906)
13. For a reproduction of the speech see: K. E. Drexler: *Nanosystems: Molecular Machinery, Manufacturing, and Computation* (Wiley-VCH, Weinheim 1992)
14. P.P. Edwards, R.L. Johnston, C.N.R. Rao: The size-induced metal–insulator transition in clusters and metal particles in: *Metal clusters in Chemistry*, ed by P. Braunstein, G. Oro, P.R. Raithbay (Wiley-VCH, Weinheim 1998)
15. A.I. Kirkland, D.E. Jefferson, D.G. Duff et al.: Proc. R. Soc. London A **440** 589 (1993)
16. T.P. Martin, T. Bergmann, H. Göhlich et al.: J. Phys. Chem. B **95** 6421 (1991)
17. B.G. Bagley: Nature **208** 674 (1965)
18. B.G. Bagley: J. Cryst. Growth **6** 323 (1970)
19. Y. Fukano, C.M. Wayman: J. Appl. Phys. **40** 1656 (1969)
20. J.G. Allpress, J.V. Sanders: Austral. J. Phys. **23** 23 (1970)
21. J.G. Allpress, J.V. Sanders: Surf. Sci. **7** 1 (1967)
22. S.H. Yang, D.A. Drabold, J.B. Adams et al.: Phys. Rev. **B47** 1567 (1993)
23. S. Ino: J. Phys. Soc. Jpn. **27** 941 (1969)
24. S. Ino: J. Phys. Soc. Jpn. **21** 346 (1966)
25. P. Chinni: Gazz. Chim. Ital. **109** 225 (1979)
26. P. Chinni: J. Organomet. Chem. **200** 37 (1980)
27. J. Kepler: *Seu De Niue Sexangula* (Tampach, 1611)
28. George F. Szprio: *Kepler's Conjecture: How Some of the Greatest Minds in History Helped Solve One of the Oldest Math Problems in the World* (Wiley, Hoboken 2003)

29. T.P. Martin: Phys. Rep. **273** 199 (1996)
30. H. Klug, L.E. Alexander: *X-ray Diffraction Procedures for Polycrystalline and Amorphous Materials* (Wiley, New York 1974)
31. D.G. Duff, A.C. Curtis, P.P. Edwards et al.: J. Chem. Soc. Chem. Commun. 1264 (1987)
32. P.A. Buffat, M. Flüeli, R. Spycher et al.: Farad. Discuss. **92** 173 (1991)
33. A.I. Kirkland, D.E. Jefferson, D. Tang et al.: Proc. R. Soc. London A **434** 279 (1991)
34. D.G. Duff, A.C. Curtis, P.P. Edwards et al.: Angew. Chem. Int. Ed. **26** 676 (1987)
35. J.O. Bovin, J.O. Malm: Z. Phys. D: Atoms, Molecules and Clusters **19** 293 (1991)
36. M.T. Reetz, W. Helbig, S.A. Quaiser et al.: Science **276** 367 (1995)
37. D.G. Duff, P.P. Edwards, J. Evans et al.: Angew. Chem. Int. Ed. **28** 590 (1989)
38. J.L. Dormann, D. Fiorani: *Magnetic Properties of Fine Particles* (North-Holland, Amsterdam 1992)
39. Q.A. Pankhurst, R.J. Pollard: J. Phys. Condens. Matter **5** 8487 (1993)
40. C. Liu, A.J. Rondinone, Z.J. Zhang: Pure Appl. Chem. **72** 37 (2000)
41. S. Chikazumi, S.H. Charap: *Physics of Magnetism* (Wiley, New York 1964)
42. B.D. Cullity: *Introduction to Magnetic Materials* (Addison-Wesley, Boston 1972)
43. L. Neel: C.R. des séances de l'Acad. des Sci. **237** 1468 (1953)
44. M. Ghosh, E.V. Sampathkumaran, C.N.R. Rao: Chem. Mater. **17** 2348 (2005)
45. M. Ghosh, K. Biswas, A. Sundaresan et al.: J. Mater. Chem. **16** 106 (2006)
46. E.D. Torre: *Magnetic Hysteresis* (IEEE, NJ 1992)
47. P.E. Jönsson: Adv. Chem. Phys. **128** 191 (2004)
48. D.G. Rancourt: Rev. Min. Geochem. **44** 217 (2001)
49. R. Kubo: J. Phys. Soc. Jpn. **17** 975 (1962)
50. S.V. Gaponenko: *Optical Properties Of Semiconductor Nanocrystals* (Cambridge University Press, Cambridge 1998)
51. W.P. Halperin: Rev. Mod. Phys. **58** 533 (1986)
52. W.A. de Heer: Rev. Mod. Phys. **65** 611 (1993)
53. U. Näher, U. Zimmermann, T.P. Martin: J. Chem. Phys. **99** 2256 (1993)
54. M. Brack: Rev. Mod. Phys. **65** 677 (1993)
55. H. Häkkinen, M. Moseler, U. Landman: Phys. Rev. Lett. **89** 033401 (2002)
56. H. Häkkinen, B. Yoon, U. Landman et al.: J. Phys. Chem. **A107** 6168 (2003)
57. W.D. Luedtke, U. Landman: J. Phys. Chem. **100** 13323 (1996)
58. U. Landman, W.D. Luedtke: Farad. Discuss. **125** 1 (2004)
59. M. Seidl, J.P. Perdew, M. Brajczewska: J. Chem. Phys. **108** 8182 (1998)
60. M.M. Kappes: Chem. Rev. **88** 369 (1988)
61. C.P. Collier, T. Vossmeyer, J.R. Heath: Annu. Rev. Phys. Chem. **49** 371 (1998)
62. *Single Electron Tunneling*, ed by H. Grabert, M.H. Devoret (Plenum, New York 1992)
63. *Nanoparticles and Nanostructured Films*, ed by J.H. Fendler (Wiley-VCH, Weinheim 1998)
64. D.L. Feldheim, C.D. Keating: Chem. Soc. Rev. **27** 1 (1998)
65. D.V. Averin, K.K. Likharev: J. Low Temp. Phys. **62** 345 (1986)
66. S. Link, M.A. El-Sayed: Int. Rev. Phy. Chem. **19** 409 (2000)
67. S. Link, M.A. El-Sayed: J. Phys. Chem. **B105** 1 (2001)

68. P. Mulvaney: Langmuir **12** 788 (1996)
69. G.C. Papavassilliou: Prog. Solid State Chem. **12** 185 (1980)
70. P.B. Johnson, R.W. Christy: Phys. Rev. **B6** 4370 (1972)
71. D.C. Look: J. Colloid Interface Sci. **56** 386 (1976)
72. http://diogenes.iwt.uni-bremen.de/~wriedt/Mie_Type_Codes/body_mie_type_codes.html
73. C.F. Bohren, D.R. Huffman: *Absorption and Scattering of Light by Small Particles* (Wiley, New York 1983)
74. M. Kerker: *The Scattering of Light and Other Electromagnetic Radiation* (Academic, New York 1969)
75. *Optical properties of metal clusters*, ed by U. Kreibig, M. Vollmer (Springer, Berlin Heidelberg New York 1995)
76. A.C. Templeton, J.J. Pietron, R.W. Murray et al.: J. Phys. Chem. **B104** 564 (2000)
77. C.A. Mirkin, R.L. Letsinger, R.C. Mucic et al.: Nature **382** 607 (1996)
78. A.L. Efros, M. Rosen: Annu. Rev. Mater. Sci. **30** 475 (2000)
79. L.E. Brus: J. Chem. Phys. **79** 5566 (1983)
80. L.E. Brus: J. Chem. Phys. **80** 4403 (1984)
81. P.E. Lippens, M. Lannoo: Phys. Rev. **B39** 10935 (1989)
82. M.V.R. Krishna, R.A. Friesner: J. Chem. Phys. **95** 8309 (1991)
83. B.O. Dabbousi, J. Rodriguez-Viejo, F.V. Mikulec et al.: J. Phys. Chem. **B101** 9463 (1997)
84. M.T. Swihart: Curr. Opin. Colloid Interface Sci. **8** 127 (2003)
85. V. Papaefthymiou, A. Kostikas, A. Simopoulos et al.: J. Appl. Phys. **67** 4487 (1990)
86. N. Saegusa, U. Kusunoli: Jpn. J. Appl. Phys. **29** 876 (1990)
87. Y.W. Du, J. Wu, H.-X. Lu et al.: J. Appl. Phys. **61** 3314 (1987)
88. C. Baker, S.K. Hasanain, S. Ismat Shah: J. Appl. Phys. **96** 6657 (2004)
89. K.-M. Lee, D.-J. Lee, H. Ahn: Mater. Lett. **58** 3122 (2004)
90. M. Agata, H. Kurase, S. Hayashi, K. Yayamoto: Solid State Commun. **76** 1061 (1990)
91. I.K. El-Zawawi, A.M. El-Shabiny: Egypt J. Solids **27** 223 (2004)
92. I. Aruna, B.R. Mehta, L.K. Malhotra et al.: Adv. Funct. Mater. **15** 131 (2005)
93. A. Maisels, F.E. Kruis, H. Fissan et al.: Appl. Phys. Lett. **77** 4431 (2000)
94. K. Nakaso, M. Shimada, K. Okuyama et al.: J. Aerosol Sci. **33** 1061 (2002)
95. P. Krasnochtchekov, R.S. Averback: J. Chem. Phys. **122** 044319 (2005)
96. A.P. Weber, M. Seipenbusch, G. Kasper: J. Phys. Chem. **A105** 8958 (2001)
97. C. Balasubramanian, V.P. Godbole, V.K. Rohatgi et al.: Nanotechnology **15** 370 (2004)
98. M. Iwata, K. Adachi, S. Furukawa et al.: J. Phys. D: Appl. Phys. **37** 1041 (2004)
99. E.F. Rexer, D.B. Wilbur, J.L. Mills et al.: Rev. Sci. Instrum. **71** 2125 (2000)
100. Y. Shimizu, K. Shiraishi, Y. Moriyoshi et al.: J. Mater. Sci. **22** 4991 (1987)
101. T. Kameyama, K. Sakanaka, A. Motoe et al.: J. Mater. Sci. **25** 1058 (1990)
102. A. Bapat, C. Anderson, C.R. Perrey et al.: Plasma Phys. Control Fusion **B97** 46 (2004)
103. S. Son, M. Taheri, E. Carpenter et al.: J. Appl. Phys. **91** 7589 (2002)
104. S. Son, R. Swaminathan, M.E. McHenry: J. Appl. Phys. **93** 7495 (2003)
105. J.-G. Li, H. Kamiyama, X.-H. Wang et al.: J. Eur. Cer. Soc. **26** 423 (2006)

106. F.K. Urban III, A. Hosseini-Tehrani, P. Griffiths et al.: J. Vac. Sci. Technol. **B20** 995 (2002)
107. D. Babonneau, A. Naudon, T. Cabioch et al.: J. Appl. Cryst. **33** 437 (2000)
108. R.C. Birtcher, S.E. Donnelly, S. Schlutig: Phys. Rev. Lett. **85** 4968 (2000)
109. S.C. Glade, T.W. Trelenberg, J.G. Tobin et al.: Mat. Res. Soc. Symp. Pro. 802 (2004)
110. T.W. Trelenberg, S.C. Glade, T.E. Felter et al.: Rev. Sci. Instrum. **75** 713 (2004)
111. J.Y. Suh, R. Lopez, L.C. Feldman et al.: J. Appl. Phys. **96** 1209 (2004)
112. S.A. Harfenist, Z.L. Wang, R.L. Whetten et al.: Adv. Mater. **9** 817 (1997)
113. S. Maruyama, L.R. Anderson, R.E. Smalley: Rev. Sci. Instrum. **61** 3686 (1990)
114. G. Raina, G.U. Kulkarni, R.T. Yadav et al.: Proc. Ind. Acad. Sci. Chem. Sci. **112** 1 (2000)
115. W. Marine, L. Patrone, B. Lukyanchuk et al.: Appl. Surf. Sci. **154–155** 345 (2000)
116. G. Ledoux, J. Gong, F. Huisken et al.: Appl. Phys. Lett. **80** 4834 (2002)
117. G. Ledoux, D. Amans, J. Gong et al.: Mater. Sci. Eng. **C19** 215 (2002)
118. M. Ehbrecht, F. Huisken: Phys. Rev. **B59** 2975 (1999)
119. X. Li, Y. He, M.T. Swihart: Langmuir **20** 4720 (2004)
120. H. Hofmeister, F. Huisken, B. Kohn: Eur. Phys. J. **D9** 137 (1999)
121. X.Q. Zhao, S. Veintemillas-Verdaguer, O. Bomati-Miguel et al.: Phys. Rev. **B71** 024106 (2005)
122. E. Borsella, S. Botti, M.C. Cesile et al.: J. Mater. Sci. Lett. **20** 187 (2001)
123. W.R. Cannon, S.C. Danforth, J.H. Flint et al.: J. Am. Ceram. Soc. **65** 324 (1982)
124. C.A. Grimes, D. Qian, E.C. Dickey et al.: J. Appl. Phys. **87** 5642 (2000)
125. R. Alexandrescu, F. Dumitrache, I. Morjan et al.: Nanotechnology **15** 537 (2004)
126. A. Watanabe, M. Unno, F. Hojo et al.: Mater. Lett. **57** 3043 (2003)
127. J.H. Kim, T.A. Germer, G.W. Mulholland et al.: Adv. Mater. **14** 518 (2002)
128. P.P. Ahonen, J. Joutsensaari, O. Richard et al.: J. Aerosol Sci. **32** 615 (2001)
129. L. Amirav, A. Amirav, E. Lifshitz: J. Phys. Chem. B **109** 9857 (2005)
130. Y.T. Didenko, K.S. Suslick: J. Am. Chem. Soc. **127** 12196 (2005)
131. L. Mädler, H.K. Kammler, R. Mueller et al.: J. Aerosol Sci. **33** 369(2002)
132. F. Miani, F. Maurigh : *Dekker Encyclopedia of Nanoscience and Nanotechnology*, p 1787–1795 (Dekker, New York 2004)
133. T.J. Colla, H.M. Urbassek: Nucl. Instrum. Methods: Phys. Res., Sect. B **164–165**, 687 (2000)
134. M. Kappes, R.W. Kunz, E. Schumacher: Chem. Phys. Lett. **91** 413 (1982)
135. S. Bandow, A.K. Kimura: Solid State Commun. **73** 167 (1990)
136. Y. Saito, T. Maki, T. Okazaki et al.: Z. Phys. **D12** 127 (1989)
137. J. Ng-Yelim, C. Roy, C. Morel: J. Mater. Sci. Lett. **8** 86 (1989)
138. Y. Yumoto, H. Shinojima, N. Uesugi et al.: Appl. Phys. Lett. **57** 2393 (1990)
139. B.G. Potter, J.H. Simmons: J. Appl. Phys. **68** 1218 (1990)
140. T. Ono, M. Kagawa, Y. Syono: J. Mater. Sci. **20** 2483 (1987)
141. C. Burda, X. Chen, R. Narayanan et al.: Chem. Rev. **105** 1025 (2005)
142. B.L. Cushing, V.L. Kolesnichenko, C.J. O'Connor: Chem. Rev. **104** 3893 (2004)
143. M.-C. Daniel, D. Astruc: Chem. Rev. **104** 293 (2004)

144. J. Turkevich, P.C. Stevenson, J. Hillier: J. Discuss. Farad. Soc. **11** 55 (1951)
145. H.I. Schlesinger, H.C. Brown, A.E. Finholt et al.: J. Am. Chem. Soc. **75** 215 (1953)
146. H.C. Brown, C.A. Brown: J. Am. Chem. Soc. **84** 1493 (1962)
147. P.R. van Rheenen, M.J. McKelvey, R. Marzke et al.: Inorg. Synth. **24** 238 (1983)
148. H. Hirai, H. Wakabayashi, M. Komiyama: Chem. Lett. 1047 (1983)
149. H. Hirai, H. Wakabayashi, M. Komiyama: Bull. Chem. Soc. Jpn. **59** 367 (1986)
150. M. Green, P. O'Brien: Chem. Commun. 1912 (2001)
151. M. Brust, M. Walker, D. Bethell et al.: J. Chem. Soc., Chem. Commun. 801 (1994)
152. R.G. Nuzzo, D.L. Allara: J. Am. Chem. Soc. **105** 4481 (1983)
153. Y. Xia, G.M. Whitesides: Angew. Chem. Int. Ed. **37** 550 (1998)
154. M.J. Hostetler, J.J. Stokes, R.W. Murray: Langmuir **12** 3604 (1996)
155. J.R. Heath, C.M. Knobler, D.V. Leff: J. Phys. Chem. **B101** 189 (1997)
156. B.A. Korgel, S. Fullam, S. Connolly et al.: J. Phys. Chem. **B102** 8379 (1998)
157. S. Chen, K. Huang, J.A. Stearns: Chem. Mater. **12** 540 (2000)
158. G.N. Glavee, K.J. Klabunde, C.M. Sorenson et al.: Langmuir **9** 162 (1993)
159. G.N. Glavee, K.J. Klabunde, C.M. Sorenson et al.: Inorg. Chem. **32** 474 (1993)
160. H. Bōnnemann, W. Brijoux, R. Brinkmann et al.: Angew. Chem. Int. Ed. **30** 1312 (1991)
161. H. Bōnnemann, W. Brijoux, R. Brinkmann et al.: J. Mol. Catal. **74** 323 (1992)
162. S. Sun, C.B. Murray: J. Appl. Phys. **85** 4325 (1999)
163. O. Margeat, C. Amiens, B. Chaudret et al.: Chem. Mater. **17** 107 (2005)
164. E.A. Hauser, J.E. Lynn: *Experiments in Colloid Chemistry*, p. 18 (McGraw-Hill, New York 1940)
165. J. Turkevich: Gold. Bull. **18** 86 (1985)
166. J. Turkevich, R.S.J. Miner, I. Okura et al.: Proc. Swedish Symp. Catal. **111** (1981)
167. D.N. Furlong, A. Launikonis, W.H.F. Sasse et al.: J. Chem. Soc., Farad. Trans. **80** 571 (1984)
168. A. Harriman, G.R. Millward, P. Nata et al.: J. Phys. Chem. **92** 1286 (1988)
169. R.S. Miner, S. Namba, J. Turkevich, in: Proceedings of the 7th international Congress on Catalysis, ed. T. Seiyama and K. Tanabe, Kodansha, Tokyo, 1981, p. 160.
170. H. Hirai, Y. Nakao, N. Toshima: J. Macromol. Sci. Chem. **A12** 1117 (1978)
171. H. Hirai: J. Macromol. Sci. Chem. **A13** 633 (1979)
172. H. Hirai, Y. Nakao, N. Toshima: J. Macromol. Sci. Chem. **A13** 727 (1979)
173. J.S. Bradley, J.M. Millar, E.W. Hill: J. Am. Chem. Soc. **113** 4016 (1991)
174. J.S. Bradley, E.W. Hill, S. Behal et al.: Chem. Mater. **4** 1234 (1992)
175. F. Porta, F. Ragaini, S. Cenini et al.: Gazz. Chim. Ital. **122** 361 (1992)
176. D. Mandler, I. Wilner: J. Phys. Chem. **91** 3600 (1987)
177. T. Teranishi, M. Miyake: Chem. Mater. **10** 54 (1998); also see J. Phys. Chem. **B101** 5774 (1997)
178. S. Ayyappan, R. Srinivasa Gopalan, G.N. Subbanna et al.: J. Mater. Res. **12** 398 (1997)
179. Y. Wang, N. Toshima: J. Phys. Chem. **B101** 5301 (1997)
180. N. Toshima, M. Harada, T. Yonezawa et al.: J. Phys. Chem. **95** 7448 (1991)
181. N. Toshima, T. Yonezawa, K. Kushihashi: J. Chem. Soc. Farad. Trans. **89** 2537 (1993)

182. H.N. Vasan, C.N.R. Rao: J. Mater. Chem. **5** 1755 (1995)
183. F. Fievet, J.P. Lagier, B. Blin et al.: Solid State Ionics **32/33** 198 (1989)
184. C. Ducamp-Sanguesa, R. Herrera-Urbina, M. Figlarz: J. Solid State Chem. **100** 272 (1992)
185. P.-Y. Silvert, R. Herrera-Urbina, N. Duvauchelle et al.: J. Mater. Chem. **6** 573 (1996)
186. F. Fivet, J.P. Lapier, M. Figlarz: Mater. Res. Soc. Bull. 29 (1989)
187. R. Seshadri, C.N.R. Rao: Mater. Res. Bull. **29** 795 (1994)
188. P. Saravanan, T.A. Jose, P.J. Thomas et al.: Bull. Mater. Sci. **24** 515 (2001)
189. X. Yan, H. Liu, K.Y. Liew: J. Mater. Chem. **11** 3387 (2001)
190. N. Toshima, Y. Wang: Chem. Lett. 1611 (1993)
191. N. Toshima, Y. Wang: Langmuir **10** 4574 (1994)
192. N. Toshima, Y. Wang: Adv. Mater. **6** 245 (1994)
193. N. Toshima, P. Lu: Chem. Lett. 729 (1996)
194. C. Roychowdhury, F. Matsumoto, P.F. Mutolo et al.: Chem. Mater. **17** 5871 (2005)
195. J.S. Bradley, E.W. Hill, C. Klien et al.: Chem. Mater. **5** 254 (1993)
196. J. Sinzig, L.J. De Jongh, H. Bönnemann et al.: Appl. Organomet. Chem. **12** 387 (1998)
197. J.L. Dai, K.L. Tsai: Farad. Disscus. **92** 45 (1991)
198. K.L. Tsai, J.L. Dai: J. Am. Chem. Soc. **113** 1650 (1991)
199. M.J. Wagner, J.L. Dye: Annu. Rev. Mater. Sci. **23** 223 (1993)
200. K.L. Tsai, J.L. Dai: Chem. Mater. **5** 540 (1993)
201. J.A. Nelson, L.H. Bennett, M.J. Wagner: J. Am. Chem. Soc. **124** 2979 (2002)
202. J.A. Nelson, L.H. Bennett, M.J. Wagner: J. Mater. Chem. **13** 857 (2003)
203. R.D. Rieke: Acc. Chem. Res. **10** 301 (1997)
204. G.L. Rochgort, R.D. Rieke: Inorg. Chem. **25** 348 (1986)
205. K.J. Klabunde, C. Mohs: Nanoparticles and nanostructural materials. in: *Chemistry of Advannced Materials - An overview*, ed by L.V. Interrante, M.J. Hampden-Smith, p. 271–318 (Wiley-VCH, New York 1998)
206. M.N. Vargaftik, I.I. Moiseev, D.I. Kochubey et al.: Farad. Discuss. **92** 13 (1991)
207. G. Schmid, R. Peifel, R. Boese et al,: Chem. Ber. **114** 3634 (1981)
208. M.R. Mucalo, R.P. Cooney: J. Chem. Soc., Chem. Commun. 94 (1989)
209. D.G. Duff, A. Baiker, P.P. Edwards: Langmuir **9** 2301 (1993)
210. M. Schulz-Dobricks, K.V. Sarathy, M. Jansen: J. Am. Chem. Soc. **127** 12816 (2005)
211. I. Pastoriza-Santos, L.M. Liz-Marzán: Langmuir **18** 2888 (2002)
212. I. Pastoriza-Santos, L.M. Liz-Marzán: Pure Appl. Chem. **72** 83 (2000)
213. P.R. van Rheenen, M.J. McKelvey, W.S. Glaunsinger: J. Solid State Chem. **67** 151 (1987)
214. P.R. van Rheenen, M.J. McKelvey, W.S. Glaunsinger: J. Solid State Chem. **67** 151 (1987)
215. N.R. Jana, X. Peng: J. Am. Chem. Soc. **125** 14280 (2003)
216. J.R. Heath: Science **258** 1131 (1992)
217. A. Kornowski, M. Giersig, R. Vogel et al.: Adv. Mater. **5** 634 (1993)
218. J.P. Wilcoxon, G.A. Samara: Appl. Phys. Lett. **74** 3164 (1999)
219. R.K. Baldwin, K.A. Pettigrew, E. Ratai: Chem. Commun. 1822 (2002)
220. J.A. Nelson, M.J. Wagner: Chem. Mater. **14** 915 (2002)
221. J.A. Nelson, E.L. Brant, M.J. Wagner: Chem. Mater. **15** 688 (2003)

222. Z. Hu, D.J.E. Ramirez, B.E.H. Cervera et al.: J. Phys. Chem. B **109** 11209 (2005)
223. G. Li, L. Li, J.B. Goated et al.: J. Am. Chem. Soc. **127** 8659 (2005)
224. M. Rajamathi, R. Seshadiri: Curr. Opn. Sol. Stat. Mater. Sci. **6** 337 (2002)
225. S.H. Yu, J. Yang, Z.H. Han et al.: J. Mater. Chem. **9** 1283 (1999)
226. X. Chen, R. Fan: Chem. Mater. **13** 802 (2001)
227. U.K. Gautam, M. Rajamathi, F. Meldrum et al.: Chem. Commun. 629 (2001)
228. D.D. Wei, Y. Jun-Sheng, P. Yi: J. Colloid Interface Sci. **299** 225 (2006)
229. L. Yang, Q. Shen, J. Zhou et al.: Mater. Chem. Phys. **98** 125 (2006)
230. W. Wang, P. Yan, F. Liu et al.: J. Mater. Chem. **8** 2321 (1998)
231. J.H. Zhan, X.G. Yang, Y. Xie et al.: J. Mater. Res. **14** 4418 (1999)
232. W. Wang, Y. Geng, Y. Qian et al.: Mater. Res. Bull. **34** 403 (1999)
233. W. Wang, Y. Geng, Y. Qian et al.: Mater. Res. Bull. **34** 131 (1999)
234. Y. Xiong, Y. Xie, G. Du et al.: J. Mater. Chem. **12** 98 (2002)
235. J. Yang, G.H. Cheng, J.H. Zeng et al.: Chem. Mater. **13** 848 (2001)
236. B. Li, Y. Xie, H. Huang et al.: Adv. Mater. **11** 1456 (1999)
237. J. Xiao, Y. Xie, Y. Xiong et al.: J. Mater. Chem. **11** 1417 (2001)
238. J. Hu, Q. Lu, K. Tang et al.: Chem. Commun. 1093 (1999)
239. Q. Lu, J. Hu, K. Tang et al.: Inorg. Chem. **39** 1606 (2000)
240. J.H. Zhan, X.G. Yang, W.X. Zhang et al.: J. Mater. Res. **15** 629 (2000)
241. Q. Peng, Y. Dong, Z. Deng et al.: Inorg. Chem. **40** 3840 (2001)
242. Y. Li, Y. Ding, Y. Qian et al.: Inorg. Chem. **37** 2844 (1998)
243. C. Zhang, Z. Kang, E. Shen et al.: J. Phys. Chem. **B110** 184 (2006)
244. J. Zu, J.-P. Ge, Y.-D. Li: J. Phys. Chem. **B110** 2497 (2006)
245. M.-O. M. Piepenbrock, T. Stirner, S.M Kelly et al.: J. Am. Chem. Soc. **128** 7087 (2006)
246. Y. Xie, Y. Qian, W. Wang et al.: Science **272** 1926 (1996)
247. K. Sardar, C.N.R. Rao: Adv. Mater. **16** 425 (2004)
248. U.K. Gautam, K. Sardar, F.L. Deepak et al.: Pramana **65** 549 (2005)
249. K. Sardar, F.L. Deepak, A. Govindaraj et al.: Small **1** 91 (2005)
250. K. Biswas, C.N.R. Rao: J. Phys. Chem. **B110** 842 (2006)
251. N. Pinna, G. Garweitner, M. Antonietti et al.: J. Am. Chem. Soc. **127** 5608 (2005)
252. W.S. Seo, J.H. Shim, S.J. Oh et al.: J. Am. Chem. Soc. **127** 6188 (2005)
253. J. Belloni, M. Mostafavi: Radiation-induced metal clusters, nucleation mechanism and chemistry. in: *Metal Clusters in Chemistry*, Vol 2, ed by P. Braunstein, L.A. Oro, P.R. Raithby (Wiley-VCH, Weinheim 1999)
254. M.Y. Han, L. Zhow, C.H. Quek et. al.: Chem. Phys. Lett. **287** 47 (1998)
255. M. Marandi, N. Taghavinia, A.I. Zad et al.: Nanotechnolgy **16** 334 (2005)
256. Z.S. Pillai, P.V. Kamat: J. Phys. Chem. **B107** 945 (2003)
257. M.T. Reetz, W. Helbig: J. Am. Chem. Soc. **116** 7401 (1994)
258. J.A. Becker, R. Schäffer, W. Festag et al.: J. Chem. Phys. **103** 2520 (1995)
259. M.T. Reetz, G. Lohmer: Chem. Commun. 1921 (1996)
260. M.T. Reetz, W. Helbig, S.A. Quaiser: Chem. Mater. **7** 2227 (1995)
261. M.T. Reetz, S.A. Quaiser: Angew. Chem. Int. Ed. **34** 2240 (1995)
262. U. Kolb, S.A. Quaiser, M. Winter et al.: Chem. Mater. **8** 1889 (1996)
263. M.T. Reetz, S.A. Quaiser, C. Merk: Chem. Ber. **129** 741 (1996)
264. M.T. Reetz, W. Helbig, S.A. Quaiser: *Active Metals*, ed by A. Fürstner, p. 339 (Wiley-VCH, Weinham 1996)

265. H. Bonnemann, R.M. Richards: Eur. J. Inorg. Chem. 2455 (2001)
266. C. Pascal, J.L. Pascal, F. Favier et al.: Chem. Mater. **11** 141 (1999)
267. R. Rossetti, J.L. Ellison, J.M. Gibson et al.: J. Chem. Phys. **80** 4464 (1984)
268. C.-H. Fischer, A. Henglein: J. Phys. Chem. **93** 5578 (1989)
269. T. Vossmeyer, L. Katsikas, M. Giersig et al.: J. Phys. Chem. **98** 7665 (1994)
270. R. Rossetti, R. Hull, J.M. Gibson: J. Chem. Phys. **82** 552 (1985)
271. A.L.P. Cornacchio, N.D. Jones: J. Mater. Chem. **16** 1171 (2006)
272. A.L. Rogach, A. Kornowski, M. Gao et al.: J. Phys. Chem. **B103** 3065 (1999)
273. A.L. Rogach, L. Katsikas, A. Kornowski et al.: Ber. Bunsenges. Phys. Chem. **100** 1772 (1996)
274. A.L. Rogach, L. Katsikas, A. Kornowski et al.: Ber. Bunsenges. Phys. Chem. **101** 1668 (1997)
275. A.L. Rogach, S.V. Kershaw, M. Burt et al.: Adv. Mater. **11** 552 (1999)
276. S.V. Kershaw, M. Burt, M. Harrison et al.: Appl. Phys. Lett. **75** 1694 (1999)
277. A. Eychmüller, A.L. Rogach: Pure Appl. Chem. **72** 179 (2000)
278. H. Weller, A. Fojtik, A. Henglein: Chem. Phys. Lett. **117** 485 (1985)
279. M. Haase, H. Weller, A. Henglein: Ber. Bunsenges. Phys. Chem. **92** 1103 (1988)
280. J.P. Chen, C.M. Sorenson, K.J. Klabunde: Phys. Rev. **B54** 9288 (1996)
281. S. Lefebure, E. Dubois, V. Cabuil et al.: J. Mater. Res. 2975 (1998)
282. L. Spanhel, M.A. Anderson: J. Am. Chem. Soc. **113** 2826 (1991)
283. D.W. Bahnemann, C. Karmann, M.R. Hoffmann: J. Phys. Chem. **91** 3789 (1987)
284. S. Sakohara, M. Ishida, M.A. Anderson: J. Phys. Chem. **B102** 10169 (1998)
285. S. Ammar, A. Helfen, N. Jouini et al.: J. Mater. Chem. **11** 186 (2001)
286. D. Carunto, Y. Remond, N.H. Chou et al.: Inorg. Chem. **41** 6137 (2002)
287. J.R. Thomas: J. Appl. Phys. **37** 2914 (1966)
288. D.P. Dinega, M.G. Bawendi: Angew. Chem. Int. Ed. **38** 1788 (1999)
289. M. Giersig, M. Hilgendorff: Colloid Surf. **202** 207 (2002)
290. V.F. Puntes, M.K. Krishnan, P. Alivisatos: Appl. Phys. Lett. **78** 2187 (2001)
291. S.L. Tripp, S.V. Putztay, A.E. Ribb et al.: J. Am. Chem. Soc. **124** 7914 (2002)
292. H. Bonnemann, W. Brijoux, R. Brinkmann et al.: Inorg. Chim. Acta **350** 617 (2003)
293. S.W. Kim, S.U. Son, S.S. Lee et al.: Chem. Commun. 2212 (2001)
294. C.B. Murray, S. Sun, H. Doyle et al.: MRS Bull. **26** 985 (2001)
295. S.J. Park, S. Kim, S. Lee et al.: J. Am. Chem. Soc. **122** 8581 (2000)
296. Y. Li, J. Liu, Y. Wang et al.: Chem. Mater. **13** 1008 (2001)
297. S. Sun, C.B. Murray, D. Weller et al.: Science **287** 1989 (2000)
298. M. Chen, D.E. Nikles: J. Appl. Phys. **91** 8477 (2002)
299. K. Ono, Y. Kakefuda, R. Okuda et al.: J. Appl. Phys. **91** 8480 (2002)
300. M. Nakamoto, M. Yamamoto, M. Fukusumi: Chem. Commun. 1622 (2002)
301. K. Philippot, B. Chaudret: C.R. Chimie **6** 1019 (2003)
302. F. Dumestre, S. Martinez, D. Zitoun et al.: Farad. Discuss. **125** 265 (2004)
303. D. de Caro, V. Agelou, A. Duteil et al.: New J. Chem. **19** 1265 (1995)
304. K. Abe, T. Hanada, Y. Yoshida et al.: Thin Solid Films **327–329** 524 (1998).
305. S.W. Kim, J. Park, Y. Jang et al.: NanoLetters **3** 1289 (2003)
306. J. Hambrock, R. Becker, A. Birkner et al.: Chem. Commun. 68 (2002)
307. X. Zhong, Y. Feng, I. Lieberwirth et al.: Chem. Mater. **18** 2468 (2006)
308. S. Sun: Adv. Mater. **18**, 393 (2006)
309. S. Kang, J.W. Harrell, D.E. Nikles: NanoLetters **2** 1033 (2002)

310. S.S. Kang, D.E. Nikles, J.W. Harrell: J. Appl. Phys. **93** 7178 (2003)
311. M. Chen, D.E. Nikles: NanoLetters **2** 211 (2002)
312. X. Sun, S. Kang, J.W. Harrell et al.: J. Appl. Phys. **93** 7337 (2003)
313. M. Chen, D.E. Nikles: J. Appl. Phys. **91** 8477 (2002)
314. K. Ono, R. Okuda, Y. Ishii et al.: J. Phys. Chem. **B107** 1941 (2003)
315. F. Dumestre, S. Martinez, D. Zitoun et al.: Farad. Discuss. **125** 265 (2003)
316. K. Ono, Y. Kakefuda, R. Okuda et al.: J. Appl. Phys. **91** 8480 (2002)
317. H. Gu, B. Xu, J. Rao et al.: J. Appl. Phys. **93** 7589 (2003)
318. M.D. Bentzon, J. van Wonterghem, S. Mørup et al.: Philos. Mag. **B60** 169 (1989)
319. T. Hyeon, S.S. Lee, J. Park et al.: J. Am. Chem. Soc. **123** 12798 (2001)
320. T. Hyeon, Y. Chung, J. Park et al.: J. Phys. Chem. **B106** 6831 (2002)
321. A.B. Bourlinos, A. Simopoulos, D. Petrides: Chem. Mater. **14** 899 (2002)
322. P.J. Thomas, P. Saravanan, G.U. Kulkarni et al.: Pramana J. Phys. **58** 371 (2002)
323. A.B. Bourlinos, A. Simopoulos, D. Petrides: Chem. Mater. **14** 2628 (2002)
324. Y. Li, M. Afzaal, P. OBrien: J. Mater. Chem. **16** 2175 (2006)
325. S. O'Brien, L. Brus, C.B. Murray: J. Am. Chem. Soc. **123** 12085 (2001)
326. C.B. Murray, D.J. Norris, M.G. Bawendi: J. Am. Chem. Soc. **115** 8706 (1993)
327. J.E. Bowen-Katari, V.L. Colvin, A.P. Alivisatos J. Phys. Chem. **98** 4109 (1994)
328. L. Qu, X. Peng: J. Am. Chem. Soc. **124** 2049 (2002)
329. Z.A. Peng, X. Peng: J. Am. Chem. Soc. **123** 183 (2001)
330. Z.A. Peng, X. Peng: J. Am. Chem. Soc. **124** 3343 (2002)
331. L. Qu, Z.A. Peng, X. Peng: NanoLetters **1** 333 (2001)
332. J. Hambrock, A. Birkner, R.A. Fischer: J. Mater. Chem. **11** 3197 (2001)
333. M.A. Hines, P. Guyot-Sionnest: J. Phys. Chem. **B102** 3655 (1998)
334. C.B. Murray, S. Sun, W. Gaschler et al.: IBM J. Res. Dev. **45** 47 (2001)
335. J. Joo, H.B. Na, T. Yu et al.: J. Am. Chem. Soc. **125** 11100 (2003)
336. A. Ghezelbash, B.A. Korgel: Langmuir **21** 9451 (2005)
337. Y.A. Yang, H. Wu, K.R. Williams et al.: Angew. Chem. Int. Ed. **44** 6712 (2005)
338. J. Jasieniak, C. Bullen, J.V. Embden et al.: J. Phys. Chem. **B109** 20665 (2005)
339. A.V. Firth, Y. Tao, D. wang, J. Ding et al.: J. Mater. Chem. **15** 4367 (2005)
340. E.M. Chan, A.P. Alivisatos, R.A. Mathies: J. Am. Chem. Soc. **127** 13854 (2005)
341. B.D. Dickerson, D.M. Irving, E. Herz et al.: Appl. Phys. Lett. **86** 171915 (2005)
342. P.D. Cozzoli, L. Manna, M.L. Curri et al.: Chem. Mater. **17** 1296 (2005)
343. J.E. Murphy, M.C. Beard, A.G. Norman et al.: J. Am. Chem. Soc. **128** 3241 (2006)
344. L. Cademartiri, J. Bertolotti, R. Sopienza et al.: J. Phys. Chem. **B110** 671 (2006)
345. T. Trindade, P. O'Brien: Adv. Mater. **8** 161 (1996)
346. T. Trindade, P. O'Brien, X. Zhang: Chem. Mater. **9** 523 (1997)
347. B. Ludolph, M.A. Malik, P. O'Brien et al.: Chem. Commun. 1849 (1998)
348. N. Revaprasadu, M.A. Malik, P. O'Brien et al.: J. Mater. Chem. **8** 1885 (1998)
349. T. Trindade, P. O'Brien, X. Zhang et al.: J. Mater. Chem. **7** 1011 (1997)
350. N. Pradhan, S. Efrima: J. Am. Chem. Soc. **125** 2050 (2003)
351. D.J. Crouch, P. O'Brien, M.A. Malik et al.: Chem. Commun. 1454 (2003)
352. Z.H. Zhang, W.S. Chin, J.J. Vittal: J. Phys. Chem. **B108** 18569 (2004)
353. Y. Hasegawa, M. Afzaal, P. O'Brien et al.: Chem. Commun. 242 (2005)

354. T. Mirkovic, M.A. Hines, P.S, Nair et al.: Chem. Mater. **17** 3451 (2005)
355. R.L. Wells, C.G. Pitt, A.T. McPhail et al.: Chem. Mater. **1** 4 (1989)
356. M.D. Healy, P.E. Laibinis, P.D. Stupik et al.: Chem. Commun. 359 (1989)
357. M.A. Olshavsky, A.B. Goldstein, A.P. Alivisatos: J. Am. Chem. Soc. **112** 9438 (1990)
358. S. Schulz, L. Martinez, J.L. Ross: Adv. Mater. Opt. Electron. **6** 185 (1996)
359. R.L. Wells, R.S. Aubuchon, S.S. Kher et al.: Chem. Mater. **7** 793 (1995)
360. O.I. Mícíc, C.J. Curtis, K.M. Jones et al.: J. Phys. Chem. **98** 4966 (1994)
361. O.I. Mícíc, J.R. Sprague, C.J. Curtis et al.: J. Phys. Chem. **99** 7754 (1995)
362. O.I. Mícíc, A.J. Nozik: J. Lumin. **70** 95 (1996)
363. A.A. Guzelian, J.E.B. Katari, A.V. Kadavanich et al.: J. Phys. Chem. **100** 7212 (1996)
364. A.A. Guzelian, U. Banin, A.V. Kadavanich et al.: Appl. Phys. Lett. **69** 1432 (1996)
365. J.F. Janik, R.L. Wells: Inorg. Chem. **36** 4135 (1997)
366. J.A. Jegier, S. McKernan, A.P. Purdy et al.: Chem. Mater. **12** 1003 (2000)
367. O.I. Mícíc, S.P. Ahrenkiel, D. Bertram et al.: Appl. Phys. Lett. **75** 478 (1999)
368. A.C. Frank, F. Stowasser, H. Sussek et al.: J. Am. Chem. Soc. **120** 3512 (1998)
369. A. Manz, A. Birkner, M. Kolbe et al.: Adv. Mater. **12** 569 (2000)
370. K. Sardar, M. Dan, B. Schwenzer et al.: J. Mater. Chem. **15** 2175 (2005)
371. S.S. Kher, R.L. Wells: Chem. Mater. **6** 2056 (1994)
372. M. Green, S. Norger, P. Moriarty et al.: J. Mater. Chem. **10** 1939 (2000)
373. M.A. Malik, P. O'Brien, M. Helliwell: J. Mater. Chem. **15** 1463 (2005)
374. C.J. Sandroff, J.P. Harbison, R. Ramesh et al.: Science **245** 391 (1989)
375. M. Green, P. O'Brien: Chem. Commun. 2459 (1998)
376. M. Green, P. O'Brien: Adv. Mater. **10** 527 (1998)
377. M. Green, P. O'Brien: J. Mater. Chem. **9** 243 (1999)
378. Z. Ding, B.M. Quinn, S.K. Haram et al.: Science **296** 1293 (1992)
379. J.D. Holmes, K.J. Ziegler, R.C. Doty et al.: J. Am. Chem. Soc. **123** 3743 (2001)
380. D.S. English, L.E. Pell, Z. Yu et al.: NanoLetters **2** 681 (2002)
381. C.L. Baigent, G.A. Muller: Experientia **36** 472 (1980)
382. M. Gutierrez, A. Henglein, J.K. Dohrmann: J. Phys. Chem. **91** 6687 (1987)
383. S.A. Yeung, R. Hobson, S. Biggs et al.: J. Chem. Soc. Chem. Commun. 378 (1993)
384. F. Grieser, R. Hobson, J.Z. Sostaric et al.: Ultrasonics **34** 547 (1996)
385. M. Ashokkumar, F. Grieser *Sonochemical Preparation of Colloids: Encyclopedia of Surface and Colloid science*, p. 4760 (Dekker, New York 2002)
386. K. Okitsu, A. Yue, S. Tanabe et al.: Langmuir **17** 7717 (2001)
387. K. Okitsu, A. Yue, S. Tanabe et al.: Bull. Chem. Soc. Jpn. **75** 2289 (2002)
388. R.A. Caruso, M. Ashokkumar, F. Grieser: Colloid Surf. **A169** 219 (2000)
389. Y. Mizukoshi, R. Oshima, Y. Maede et al.: Langmuir **15** 2733 (1999)
390. T. Fujimoto, S. Terauchi, H. Umehara et al.: Chem. Mater. **13** 1057 (2001)
391. Y. Mizukoshi, E. Takagi, H. Okuno et al.: Ultrason. Sonochem. **8** 1 (2001)
392. T. Fujimoto, Y. Mizukoshi, Y. Nagata et al.: Scr. Mater. **44** 2183 (2001)
393. K. Okitsu, S.'Nagaoka, S. Tanabe et al.: Chem. Lett. 271 (1999)
394. Y. Nagata, S. Watananabe, S.-I. Fujita et al.: J. Chem. Soc. Chem. Commun. 1620 (1992)
395. R.A. Salkar, P. Jeevanandam, S.T. Aruna et al.: J. Mater. Chem. **9** 1333 (1999)
396. N.A. Dhas, C.P. Raj, A. Gedanken: Chem. Mater. **10** 1446 (1998)

397. K.S. Suslick, M. Fang, T. Hyeon: J. Am. Chem. Soc. **118** 11960 (1996)
398. K.V.P.M. Shafi, S. Wizel, T. Prozorov et al.: Thin solid films **318** 38 (1998)
399. Y. Koltypin, G. Katabi, X. Cao et al.: J. Non-Cryst. solids **201** 159 (1996)
400. K.V.P.M. Shafi, A. Gedanken, R. Prozorov: Adv. Mater. **10** 590 (1998)
401. N.A. Dhas, A. Gedanken: J. Mater. Chem. **8** 445 (1998)
402. N.A. Dhas, C.P. Raj, A. Gedanken: Chem. Mater. **10** 3278 (1998)
403. J.L. Heinrich, C.L. Curtis, G.M. Credo et al.: Science **255** 66 (1992)
404. R.A. Bley, S.M. Kauzlarich, J.E. Davis et al.: Chem. Mater. **8** 1881 (1996)
405. R.A. Hobson, P. Mulvaney, F Grieser: J. Chem. Soc. Chem. Commun. 823 (1994)
406. B. Li, Y. Xie, J. Huang et al.: Ultrason. Sonochem. **6** 217 (1999)
407. N.A. Dhas, K.S. Suslick: J. Am. Chem. Soc. **127** 2368 (2005)
408. K.V.P.M. Shafi, I. Felner, Y. Mastai et al.: J. Phys. Chem. **B103** 3358 (1999)
409. B. Li, Y. Xe, J.X. Huang et al.: Ultrason. Sonochem. **8** 231 (2001)
410. M.-P. Pileni: Nat. Mater. **2** 145 (2003)
411. M.-P. Pileni: J. Phys. Chem. **97** 6961 (1993)
412. M.-P. Pileni: Langmuir **13** 3266 (1997)
413. M.-P. Pileni: Langmuir **17** 7476 (2001)
414. M. Boutonnet, J. Kizling, P. Stenius: Colloid Surf. **5** 209 (1982)
415. C. Petit, T.K. Jain, F. Billoudet et al.: Langmuir **10** 4446 1994
416. P. Lianos, J.K. Thomas: Chem. Phys. Lett. **125** 299 (1986)
417. C.R. Vestal, Z.J. Zhang: J. Am. Chem. Soc. **125** 9828 (2003)
418. P.S. Shah, S. Husain, K.P. Johnston et al.: J. Phys. Chem. **B105** 9433 (2001)
419. J.D. Holmes, D.M. Lyons, K.J. Ziegler: Chem. Eur. J. **9** 2144 (2003)
420. J.P. Wilcoxon, R.L. Williamson, R. Baughman: J. Chem. Phys. **98** 9933 (1993)
421. J.P. Wilcoxon, G.A. Samara, P.N. Provencio: Phys. Rev. **B60** 2704 (1999)
422. J.P. Wilcoxon, P.P. Provencio, G.A. Samara : Phys. Rev. **B64** 64 035417 (2001)
423. N. Moumen, M.-P. Pileni: J. Phys. Chem. **100** 1867 (1996)
424. N. Moumen, M.-P. Pileni: Chem. Mater. **8** 1128 (1996)
425. S. Qiu, J. Dong, G. Chen: J Colloid Interface Sci. **216** 230 (1999)
426. S. Mandal, S.K. Arumugam, S.D. Adyanthaya et al.: J. Mater. Chem. **14** 43 (2004)
427. C.N.R. Rao, G.U. Kulkarni, P.J. Thomas et al.: Curr. Sci. **85** 1041 (2003)
428. C.N.R. Rao, G.U. Kulkarni, P.J. Thomas et al.: J. Phys. Chem. **B107** 7391 (2003)
429. C.N.R. Rao, G.U. Kulkarni, V.V. Agrawal et al.: J. Colloid Interface Sci. **289** 305 (2005)
430. U.K. Gautam, M. Ghosh, C.N.R. Rao: Langmuir **20** 10775 (2004)
431. V.V. Agrawal, G.U. Kulkarni, C.N.R. Rao: J. Phys. Chem. **B109** 7300 (2005)
432. V.V. Agrawal, P. Mahalakshmi, G.U. Kulkarni et al.: Langmuir **22** 1846 (2006)
433. J.E. Zumberg, A.C. Sieglo, B. Nagy: Miner. Sci. Eng. **10** 223 (1978)
434. M. Hosea, B. Greene, R. McPherson et al.: Inorg. Chim. Acta. **123** 161 (1986)
435. T.J. Beveridge, R.J. Doyle: *Metal ions and Bacteria* (Wiley, New York 1989)
436. H. Aiking, K. Kok, H. van Heerikhuizen et al.: Appl. Environ. Microbiol. **44** 938 (1982)
437. R.N. Reese, D.R. Winge: J. Biol. Chem. **263** 12832 (1988)
438. C.T. Dameron, R.N. Reese, R.K. Mehra et al.: Nature **338** 596 (1989)
439. K.L. Temple, N. LeRoux: Econ. Geol. **59** 647 (1964)
440. R.P. Lakemore, D. Maratea, R.S. Wolfe: J. Bacteriol. **140** 720 (1979)

441. M. Sastry, A. Ahmad, M.I. Khan et al.: Curr. Sci. **85** 170 (2003)
442. M. Sastry, A. Ahmad, M.I. Khan et al.: *Microbial Nanoparticle Production* in Nanobiotechnology, ed by C. Niemeyer and C. Mirkin (Wiley-VCH, Weinham 2003)
443. B. Nair, T. Pradeep: Cryst. Growth Des. **2** 293 (2002)
444. T. Klaus, R. Joerger, E. Olsson et al.: Proc. Natl. Acad. Sci. USA **96** 13611 (1999)
445. T. Klaus-Joerger, R. Joerger, E. Olsson et al.: Trends Biotechnol. **19** 15 (2001)
446. R. Joerger, T. Klaus, C.-G. Granqvist: Adv. Mater. **12** 407 (2000)
447. J.D. Holmes, P.R. Smith, R. Evans-Gowing et al.: Arch. Microbiol. **163** 143 (1995)
448. Y. Roh, R.J. Lauf, A.D. McMillan et al.: Solid State Commun. **118** 529 (2001)
449. I. Willner, R. Baron, B. Willner: Adv. Mater. **18** 1109 (2006)
450. M. Kowshik, W. Vogel, J. Urban et al.: Adv. Mater. **14** 815 (2002)
451. P. Mukherjee, S. Senapathi, D. Mandal et al.: ChemBioChem **3** 461 (2002)
452. A. Ahmad, P. Mukherjee, S. Senapathi et al.: Colloid Surf. B **28** 313 (2003)
453. P. Mukherjee, A. Ahmad, D. Mandal et al.: Angew. Chem. Int. Ed. **40** 3585 (2001)
454. P. Mukherjee, A. Ahmad, D. Mandal et al.: NanoLetters **1** 515 (2001)
455. A. Ahmad, P. Mukherjee, D. Mandal: J. Am. Chem. Soc. **124** 12108 (2002)
456. H. Qian, C. Dong, J. Weng et al.: Small **2** 747 (2006)
457. R.S. Bowles, J.J. Kolstad, J.M. Calo et al.: Surf. Sci. **106** 117 (1981)
458. S. Stoeva, K.J. Klabunde, C.M. Sorenson et al.: J. Am. Chem. Soc. **124** 2305 (2002)
459. J. Neddersen, G. Chumanov, T.M. Cotton: Appl. Spectrosc. **47** 1959 (1993)
460. F. Mafune, J.-y. Kohno, Y. Takeda et al.: J. Phys. Chem. **B106** 7575 (2002)
461. F. Mafune, J.-y. Kohno, Y. Takeda et al.: J. Phys. Chem. **B104** 8333 (2000)
462. F. Mafune, J.-y. Kohno, Y. Takeda et al.: J. Phys. Chem. **B104** 9111 (2000)
463. F. Mafune, J.-y. Kohno, Y. Takeda et al.: J. Phys. Chem. **B107** 4218 (2003)
464. C.J. Murphy, N.R. Jana: Adv. Mater. **14** 80 (2002)
465. N.R. Jana, L. Gearheart, C.J. Murphy: Chem. Commun. 617 (2001)
466. N.R. Jana, L. Gearheart, C.J. Murphy: J. Phys. Chem. **B105** 4065 (2001)
467. J. Gao, C.M. Bender, C.J. Murphy: Langmuir **19** 9065 (2003)
468. N.R. Jana, L. Gearheart, C.J. Murphy: Adv. Mater. **13** 1389 (2001)
469. T.K. Sahu, C.J. Murphy: J. Am. Chem. Soc. **126** 8648 (2004)
470. S. Chen, D.L. Carroll: J. Phys. Chem. **B108** 5500 (2004)
471. K.K. Caswell, C.M. Bender, C.J. Murphy: NanoLetters **3** 667 (2003)
472. Y. Sun, B. Gates, B. Mayers et al.: NanoLetters **2** 165 (2002)
473. Y. Sun, B. Mayers, T. Herricks et al.: NanoLetters **3** 955 (2003)
474. Y. Sun, Y. Xia: Science **298** 2176 (2002)
475. S.S. Shankar, A. Rai, B. Ankamwar et al.: Nature Mater. **3** 482 (2004)
476. R.K. Naik, S.J. Stringer, G. Agarwal et al.: Nature Mater. **1** 169 (2002)
477. R. Jin, Y. Cao, C.A. Mirkin: Science **294** 1901 (2001)
478. R. Jin, Y. Cao, E. Hao et al.: Nature **425** 487 (2003)
479. G.S. Metraux, C.A. Mirkin: Adv. Mater. **17** 412 (2005)
480. E. Hao, R.C. Bailey, G.C. Schatz: NanoLetters **4** 327 (2004)
481. C-H Kuo, M.H. Huang: Langmuir **21** 2012 (2005)
482. Y. Xiao, B. Shylahovsky, I. Popov et al.: Langmuir **21** 5659 (2005)
483. M. Maillard, S. Giorgio, M.-P. Pileni: Adv. Mater. **14** 1084 (2002)

484. I. Pastoriza-Santos, L.L. Marzan: NanoLetters **2** 903 (2002)
485. D.O. Yener, J. Sindel, C.A. Randell et al.: Langmuir **18** 8692 (2002)
486. Y. Shao, Y. Jin, S. Dong: Chem. Commun. 1104 (2004)
487. F. Dumestre, B. Chaudret, C. Amiens et al.: Angew. Chem. Int. Ed. **41** 4286 (2002)
488. K. Soulantica, A. Maisonnat, F. Senocq et al.: Angew. Chem. Int. Ed. **40** 2984 (2001)
489. H. Song, F. Kim, S. Connor: J. Phys. Chem. **B109** 188 (2005)
490. S.I. Stoeva, V. Zaikoski, B.L.V. Prasad et al.: Langmuir **21** 10280 (2005)
491. L. Manna, E.C. Sher, A.P. Alivisatos: J. Am. Chem. Soc. **122** 12700 (2000)
492. X. Peng, L. Manna, W. Yang et al.: Nature **404** 59 (2000)
493. Q. Pang, L. Zhao, Y. Cai et al.: Chem. Mater. **17** 5263 (2005)
494. S.D. Bunge, K.M. Krueger, R. H-Boyle et al.: J. Mater. Chem. **13** 1705 (2003)
495. L. Manna, D.J. Milliron, A. Meisel et al.: Nature Mater. **2** 382 (2003)
496. H.Q. Yan, R.R. He, J. Pham et al.: Adv. Mater. **15** 402 (2003)
497. Z. Chen, Z. Shan, M.S. Cao et al.: Nanotechnology **15** 365 (2004)
498. D.J. Milliron, S.M. Hughers, Y. Cui et al.: Nature **430** 190 (2004)
499. Y. Li, X. Li, C. Yang et al.: J. Mater. Chem. **13** 2641 (2003)
500. A.G. Kanaras, C. Sonnichsen, H. Liu et al.: NanoLetters **5** 2164 (2005)
501. Y. Cheng, Y. Wang, F. Bao et al.: J. Phys. Chem. **B110** 9448 (2006)
502. C. Qian, F. Kim, L. Ma et al.: J. Am. Chem. Soc. **126** 1195 (2004)
503. N. Pradhan, S. Efrima: J. Phys. Chem. **B108** 11964 (2004)
504. F.X. Redl, C.T. Black, G.C. Papaefthymiou et al.: J. Am. Chem. Soc. **126** 14583 (2004)
505. A.J. Houtepen, R. Koole, D. Vanmaekelbergh et al.: J. Am. Chem. Soc. **128** 6792 (2006)
506. T.H. Larsen, M. Sigman, A. Ghezelbash et al.: J. Am. Chem. Soc. **125** 5636 (2003)
507. M.B. Sigman, A. Ghezelbash, T.H. Hanrath et al.: J. Am. Chem. Soc. **125** 16050 (2003)
508. A. Ghezelbash, M.B. Sigman, B.A. Korgel et al.: NanoLetters **4** 537 (2004)
509. J. Tanori, M.-P. Pileni: Langmuir **13** 639 (1997)
510. L.M. Qi, J. Ma, H. Chen et al.: J. Phys. Chem. **B101** 340 (1997)
511. J.D. Hopwood, S. Mann: Chem. Mater. **9** 1819 (1997)
512. N. Pinna, K. Weiss, H. Sach-Kongehl et al.: Langmuir **17** 7982 (2001)
513. M. Li, H. Schnablegger, S. Mann: Nature **402** 393 (1999)
514. R.K. Baldwin, K.P. Pettigrew, J. C. Garno et al.: J. Am. Chem. Soc. **124** 1150 (2002)
515. M. Monge, M.L. Kahn, A. Maisonnat et al.: Angew. Chem. Int. Ed. **42** 5321 (2003)
516. J. Joo, S.G. Kwon, J.H. Yu et al.: Adv. Mater. **17** 1873 (2005)
517. S.-H. Choi, E.-G. Kim, J. Park et al.: J. Phys. Chem. B **109** 14792 (2005)
518. L. Levy, J.F. Hochepied, M.-P. Pileni: J. Phys. Chem. **100** 18322 (1996)
519. G. Counio, S. Esnouf, T. Gacoin et al.: J. Phys. Chem. **100** 20021 (1996)
520. D.M. Hoffman, B.K. Meyer, A.I. Ekimov et al.: Solid State Commun. **114** 547 (2000)
521. Y. Wang, N. Herron, K. Moller et al.: Solid State Commun. **77** 33 (1991)
522. R.N. Bhargava, D. Gallagher, X. Hong et al.: Phys. Rev. Lett. **72** 416 (1994)
523. K. Sooklal, B.S. Cullum, S.M. Angel et al.: J. Phys. Chem. **100** 4551 (1996)

524. F.V. Mikulec, M. Kuno, M. Bennati et al.: J. Am. Chem. Soc. **122** 2532 (2000)
525. D.J. Norris, N. Yao, F.T. Charmok et al.: NanoLetters **1** 3 (2001)
526. J.F. Suyver, S.F. Wuister, J.J. Kelly et al.: Phys. Chem. Chem. Phys. **2** 5445 (2000)
527. S.C. Erwin, L. Zu, M.I. Haftel et al.: Nature **436** 91 (2005)
528. M. Shim, C. Wang, D.J. Norris et al.: MRS Bull. **1005** (2001)
529. M. Shim, C. Wang, P. Guyot-Sionnest.: J. Phys. Chem. **B105** 2369 (2001)
530. C. Wang, M. Shim, P. Guyot-Sionnest.: Science **291** 2390 (2001)
531. W. Schartl: Adv. Mater. **12** 1899 (2000)
532. H. Harai, H. Aizawa, H. Shiozaki: Chem. Lett. **8** 1527 (1992)
533. G. Schmid: Polyhedron **7** 2321 (1988)
534. L.O. Brown, J.E. Hutchison: J. Am. Chem. Soc. **121** 882 (1999)
535. W.M. Pankau, K. Verbist, G.v. Kiedrowski: Chem. Commun. 519 (2001)
536. K.V. Sarathy, G. Raina, R.T. Yadav et al.: J. Phys. Chem. **B101** 9876 (1998)
537. K.V. Sarathy, G.U. Kulkarni, C.N.R. Rao: Chem. Commun. 537 (1997)
538. A. Kumar, A.B. Mandale, M. Sastry: Langmuir **16** 9229 (2000)
539. D.I. Gittins, F. Caruso: Angew. Chem. Int. Ed. **40** 3001 (2001)
540. D.I. Gittins, F. Caruso: Chemphyschem. **3** 110 (2002)
541. T. Pellegrino, L. Manna, S. Kudera et al.: NanoLetters **4** 703 (2004)
542. M. Bruchez Jr, M. Moronne, P. Gin et al.: Science **281** 2013 (1998)
543. W.C.W. Chan, S. Nie: Science **281** 2016 (1998)
544. S. Pathak, S.-K. Choi, N. Arnheim et al.: J. Am. Chem. Soc. **123** 4103 (2001)
545. H. Mattoussi, J.M. Mauro, E.R. Goldman et al.: J. Am. Chem. Soc. **122** 12142 (2000)
546. W.C.W. Chan, D.J. Maxwell, X. Gao et al.: Curr. Opn. Biotechnol. **13** 40 (2002)
547. W.J. Parak, D. Gerion, D. Zanchet et al.: Chem. Mater. **14** 2113 (2002)
548. K.S. Mayya, B. Scheeler, F. Caruso: Adv. Func. Mater. **13** 183 (2003)
549. T. Cassagneau, J.H. Fendler: J. Phys. Chem. **B101** 1789 (1999)
550. T. Tsukatani, H. Fujihara: Langmuir **21** 12093 (2005)
551. R. Vaggu, K. Biswas, C.N.R. Rao: J. Phys. Chem. **B** (2006)
552. G.L. Hornayak, M. Kröll, R. Pugin et al.: Eur. J. Chem. **3** 1951 (1997)
553. S. Lin, M. Li, E. Dujardin et al.: Adv. Mater. **17** 2553 (2005)
554. A.P. Alivisatos, K.P. Johnsson, X. Peng et al.: Nature **382** 609 (1996)
555. M.G. Warner, J. Hutchison: Nature Mater. **2** 272 (2003)
556. A. Kumar, M. Pattarkine, M. Bhadbhade et al.: Adv. Mater. **13** 341 (2001)
557. J. Yang, J.Y. Lee, H.-P. Too et al.: Chem. Phys. **323** 304 (2006)
558. M. Mitov, C. Portet, C. Bourgerette et al.: Nat. Mater. **1** 229 (2002)
559. S.-W. Lee, C. Mao, C.E. Flynn et al.: Science **296** 892 (2002)
560. M. Kogiso, K. Yoshida, K. Yase et al.: Chem. Commun. 2492 (2002)
561. D. Wyra, N. Beyer, G. Schmid: NanoLetters **2** 419 (2002)
562. E. Dujardin, C. Peet, G. Stubbs et al.: NanoLetters **3** 413 (2003)
563. T. Oku, K. Suganuma: Chem. Commun. 2355 (1999)
564. S.W. Chung, G. Markovich, J.R. Heath: J. Phys. Chem. **B102** 6685 (1998)
565. K.G. Thomas, S. Barazzouk, B.I. Ipe et al.: J. Phys. Chem. **B108** 13066 (2004)
566. C.N.R. Rao, A. Govindaraj, F.L. Deepak et al.: Appl. Phys. Lett. **78** 1853 (2001)
567. E.J.H. Lee, C. Ribeiro. E. Longo et al.: J. Phys. Chem. B **109** 20842 (2005)

568. P.C. Ohara, J.R. Heath, W.M. Gelbart: Angew. Chem. Int. Ed. **36** 1077 (1997)
569. T. Vossmeyer, S.-W. Chung, W.M. Gelbart et al.: Adv. Mater. **10** 351 (1998)
570. P.C. Ohara and W.M. Gelbart: Langmuir **14** 3418 (1998)
571. M. Maillard, L. Motte, A.T. Ngo et al.: J. Phys. Chem. **B104** 11871 (2000)
572. M. Maillard, L. Motte, M.P. Pileni: Adv. Mater. **16** 200 (2001)
573. C. Stowell, B.A. Korgel: NanoLetters **1** 595 (2001)
574. L.V. Govor, G.H. Bauer, G. Reiter et al.: Langmuir **19** 9573 (2003)
575. B. Liu, H.C. Zeng: J. Am. Chem. Soc. **127** 18262 (2005)
576. Z. Liu, R. Levicky: Nanotecnology **15** 1483 (2004)
577. J. Huang, F. Kim, A.R. Tao et al.: Nat. Mater. **4** 896 (2005)
578. S.A. Claridge, S.L. Goh, J.M.J. Frechet et al.: Chem. Mater. **17** 1628 (2005)
579. J.H. Warner, R.D. Tilley: Adv. Mater. **17** 2997 (2005)
580. Y. Lin, A. Boker, J. He et al.: Nature **434** 55 (2005)
581. R.L. Whetten, J.T. Khoury, M.M. Alvarez et al.: Adv. Mater. **8** 428 (1996)
582. S.I. Stoeva, B.L.V. Prasad, S. Uma et al.: J. Phys. Chem. **B107** 7441 (2003)
583. G. Schmid, M. Bäumle, N. Beyer: Angew. Chem. Int. Ed. **39** 181 (2000)
584. L.O. Brown, J.E. Hutchison: J. Phys. Chem. **B105** 8911 (2001)
585. P.J. Thomas, G.U. Kulkarni, C.N.R. Rao et al.: J. Phys. Chem. **B104** 8138 (2000)
586. S.-Y. Zhao, S. Wang, K. Kimura: Langmuir **20** 1977 (2004)
587. N. Sandhyarani, M.R. Reshmi, R. Unnikrishnan et al.: Chem. Mater. **12** 104 (2000)
588. S.T. He, S.S. Xie, J.N. Yao et al.: Appl. Phys. Lett. **81** 150 (2002)
589. C. Desvaux, C. Amiens, P. Fejes et al.: Nat. Mater. **4** 750 (2005)
590. C. Petit, A. Taleb, M.P. Pileni.: J. Phys. Chem. **B103** 1805 (1999)
591. M.P. Pileni: New. J. Chem. 696 (1998)
592. E.V. Shevchenko, D.V. Talapin, A.L. Rogach et al.: J. Am. Chem. Soc. **124** 11480 (2002)
593. H. Fan, E. Leve, J. Gabaldon et al.: Adv. Mater. **17** 2587 (2005)
594. B. Kim, S.L. Tripp, A. Wei: J. Am. Chem. Soc. **123** 7955 (2001)
595. B.A. Korgel, D. Fitzmaurice: Adv. Mater. **10** 661 (1998)
596. F. Dumestre, B. Chaudret, C. Amiens et al.: Science **303** 821 (2004)
597. K. Soulantica, A. Maisonnat, M.-C. Fromen et al.: Angew. Chem. Int. Ed. **42** 1945 (2003)
598. C.J. Kiely, J. Fink, M. Brust et al.: Nature **396** 444 (1998)
599. C.J. Kiely, J. Fink, J.G. Zheng et al.: Adv. Mater. **12** 640 (2000)
600. A.E. Saunders, B.A. Korgel: Chemphyschem. **6** 61 (2005)
601. C.B. Murray, C.R. Kagan, M.G. Bawendi: Science **270** 1335 (1995)
602. C.B. Murray, C.R. Kagan, M.G. Bawendi: Annu. Rev. Mater. Sci. **30** 545 (2000)
603. L. Motte, F. Billoudet, D. Thiaudiére et al.: J. de Phys. III **7** 517 (1997)
604. O.I. Micic, K.M. Jones, A. Cahill et al.: J. Phys. Chem. **B102** 9791 (1998)
605. Z. Liu, J. Liang, D. Xu et al.: Chem. Commun. 2724 (2004)
606. S. Sun, H. Zeng: J. Am. Chem. Soc. **124** 8204 (2002)
607. J.S. Yin, Z.L. Wang: Phys. Rev. Lett. **79** 2570 (1997)
608. F.X. Redl, K.-S. Cho, C.B. Murray et al.: Nature **423** 968 (2003)
609. E.V. Shevchenko, D.V. Talapin, C.B. Murray et al.: J. Am. Chem. Soc. **128** 3620 (2006)
610. R.A. Mcmillan, C.D. Paavola, J. Howard et al.: Nat. Mater. **1** 247 (2002)

611. L.F. Chi, S. Rakers, C.P. Daghlian et al.: Thin Solid Films **327** 520 (1998)
612. M. Sastry, A. Gole, V. Patil: Thin Solid Films **384** 125 (2001)
613. E.S. Smotkin, C. Lee, A.J. Bard et al.: Chem. Phys. Lett. **152** 265 (1988)
614. Y.S. Kang, D.K. Lee, C.S. Lee et al.: J. Phys. Chem. **B106** 9341 (2002)
615. B.A. Korgel, D. Fritzmaurice: Phys. Rev. Lett. **80** 3531 (1998)
616. T.P. Bigioni, X.-M. Lin, T.T. Nguyen et al.: Nature Mater. **5** 265 (2006)
617. M. Brust, D. Bethell, C.J. Kiely et al.: Langmuir **14** 5425 (1998)
618. T. Torimoto, N. Tsumura, M. Miyake et al.: Langmuir **15** 1853 (1999)
619. K.V. Sarathy, P.J. Thomas, G.U. Kulkarni et al.: J. Phys. Chem. **B103** 399 (1999)
620. J. Schmitt, G. Decher, W.J. Dressick et al.: Adv. Mater. **9** 61 (1997)
621. J. Schmitt, P. Mächtle, D. Eck et al.: Langmuir **15** 3256 (1999)
622. N.A. Kotov, I. Dékány, J.H. Fendler.: J. Phys. Chem. **99** 13065 (1995)
623. Y.-C. Liao, J.T. Roberts.: J. Am. Chem. Soc. **128** 9061 (2006)
624. J.J. Urban, D.V. Talapin, E.V. Shevchenko et al.:J. Am. Chem. Soc. **128** 3248 (2006)
625. R. Blonder, L. Sheeney, I. Willner: Chem. Commun. 1393 (1998)
626. M. Lahav, A.N. Shipway, I. Willner et al.: J. Electroanal. Chem. **482** 217 (2000)
627. M. Lahav, A.N. Shipway, I. Willner: J. Chem. Soc., Perkin Trans. **2** 1925 (1999)
628. M. Lahav, R. Gabai, A.N. Shipway et al.: Chem. Commun. 1937 (1999)
629. E. Hao, T. Lian: Chem. Mater. **12** 3392 (2000)
630. J.F. Hicks, Y. Seok-Shon, R.W. Murray: Langmuir **18** 2288 (2002)
631. H.G. Fritsche, H. Muller, B. Fehrensen: Z. Phy. Chem. **199** 87 (1997)
632. H. Feld, A. Leute, D. Rading et al.: J. Am. Chem. Soc. **112** 8166 (1990)
633. P.J. Thomas, G.U. Kulkarni, C.N.R. Rao: J. Phys. Chem. **B105** 2515 (2001)
634. G. Schmid, R. Pugin, T. Sawitowski et al.: Chem. Commun. 1303 (1999)
635. S. Wang, S. Sato, K. Kimura: Langmuir **15** 2445 (2003)
636. E. Shevchenko, D. Talapin, A. Kornowski et al.: Adv. Mater. **14** 287 (2002)
637. D.V. Talapin, E.V. Shevchenko, A. Kornowski et al.: Adv. Mater. **13** 1868 (2001)
638. J. Wang, H.L. Duan, Z.P. Huang et al.: Proc. R. Soc. A **462** 1355 (2006)
639. W. Thomson: Philos. Mag. **42** 448 (1871)
640. P. Pawlow: Z. Phys. Chem. **65** 545 (1909)
641. M. Takagi: J. Phys. Soc. Jpn. **9** 359 (1954)
642. Ph. Buffat, J.-P. Borel: Phys. Rev. **A13** 2287 (1976)
643. F.G. Shi: J. Mater. Res. **9** 1307 (1994)
644. Q. Jiang, S. Zhang, M. Zhao: Mater. Chem. Phys. **82** 225 (2003)
645. G.A. Breaux, R.C. Benirschke, T. Sugai et al.: Phys. Rev. Lett. **91** 215508 (2003)
646. G.A. Breaux, D.A. Hillman, C.M. Neal et al.: J. Am. Chem. Soc. **126** 8628 (2004)
647. M. Schmidt, R. Kusche, W. Kronmüller et al.: Phys. Rev. Lett. **79** 99 (1997)
648. M. Schmidt, R. Kusche, B. von Issendorf et al.: Nature **393** 238 (1998)
649. M. Schmidt, H. Haberland.: C.R. Phys. **3** 327 (2002)
650. S. Chacko, K. Joshi, D.G. Kanhere et al.: Phys. Rev. Lett. **92** 135506 (2004)
651. H. Haberland, T. Hippler, J. Donges et al.: Phys. Rev. Lett. **94** 035701 (2005)
652. J. Rupp, R. Birringer: Phys. Rev. **B36** 7888 (1987)
653. H. Frölich: Physica **4** 406 (1937)
654. H.Y. Bai, J.L. Luo, D. Jin et al.: J. Appl. Phys. **79** 361 (1996)

655. Y. Volokitin, J. Sinzig, L.J. de Jongh et al.: Nature **384** 621 (1996)
656. K. Clemenger: Phys. Rev. **B44** 12991 (1991)
657. R. Busani, M. Folker, O. Chesnovsky: Phys. Rev. Lett. **81** 3836 (1998)
658. K. Rademann, O.D. Rademann, M. Schlauf et al.: Phys. Rev. Lett. **69** 3208 (1992)
659. O. Chesnovsky, S.H. Yan, C.L. Pettiette et al.: Chem. Phys. Lett. **138** 119 (1987)
660. Y. Negishi, H. Kawamata, T. Hayase et al.: Chem. Phys. Lett. **269** 2460 (1997)
661. H.N. Aiyer, V. Vijayakrishnan, G.N. Subbanna et al.: Surf. Sci. **313** 392 (1994)
662. C.N.R. Rao, G.U. Kulkarni, P.J. Thomas et al.: Chem. Eur. J. **8** 29 (2002)
663. P.M. Paulus, A. Goossens, R.C. Thiel et al.: Phys. Rev. **B64** 205418 (2001)
664. J. Nanda, B.A. Kuruvilla, D.D. Sarma: Phys. Rev. **B59** 7473 (1999)
665. M. Rosenblit and J. Jortner: J. Phys. Chem. **98** 9365 (1994)
666. R.A. Perez, A.F. Ramos, G.L. Malli: Phys. Rev. **39** 3005 (1989)
667. O.D. Haberlen, S.C. Chung, M. Stener et al.: J. Chem. Phys. **106** 5189 (1997)
668. C.P. Vinod, G.U. Kulkarni, C.N.R. Rao: Chem. Phys. Lett. **289** 329 (1998)
669. B. Marsen, M. Lonfat, P. Scheier et al.: Phys. Rev. **B62** 6892 (2000)
670. S. Ogawa, F.F. Fan, A.J. Bard: J. Phys. Chem. **99** 11182 (1995)
671. O. Millo, D. Katz, Y. Cao et al.: Phys. Rev. **61** 16773 (2000)
672. B. Wang, X. Xiao, X. Huang et al.: Appl. Phys. Lett. **77** 1180 (2000)
673. B. Alperson, S. Cohen, I. Rubinstein et al.: Phys. Rev. **52** R17017(1995)
674. U. Banin, O. Millo: Annu. Rev. Phys. Chem. **54** 465 (2003)
675. J. Lambe, R. Jaklevic: Phys. Rev. Lett. **22** 1371 (1969)
676. M. Amman, R. Wilkins, E. Ben-Jacob et al.: Phys. Rev. **B43** 1146 (1991)
677. U. Simon: Adv. Mater. **10** 1487 (1998)
678. P.J. Thomas, G.U. Kulkarni, C.N.R. Rao: Chem. Phys. Lett. **321** 163 (2000)
679. J. Jortner: Z. Phys. D: Atoms, Molecules and Clusters **24** 247 (1992)
680. R.P. Andres, T. Bein, M. Dorogi et al.: Science **272** 1323 (1996)
681. D.L. Feldheim, K.C. Grabar, M.J. Natan et al.: J. Am. Chem. Soc. **118** 7640 (1996)
682. R.S. Ingram, M.J. Hostetler, R.W. Murray et al.: J. Am. Chem. Soc. **119** 9279 (1997)
683. S. Chen, R.S. Ingram, M.J. Hostetler et al.: Science **280** 2098 (1998)
684. S. Chen, R.W. Murray, S.W. Feldberg: J. Phys. Chem. **B102** 9898 (1998)
685. S. Chen, R.W. Murray: Langmuir **3** 682 (1998)
686. M. Brust, D. Bethell, D.J. Schiffrin et al.: Adv. Mater. **7** 795 (1995)
687. V. Torma, G. Schmid, U. Simon.: Chemphyschem. **2** 321 (2001)
688. U. Simon, R. Flesch, H. Wiggers et al.: J. Mater. Chem. **8** 517 (1998)
689. M. Aslam, I.S. Mulla, K. Vijayamohanan: Appl. Phys. Lett. **79** 689 (2001)
690. H.E. Romero, M. Drndic: Phys. Rev. Lett. **95** 156801 (2005)
691. R.H. Terrill, T.A. Postlewaite, C. Chen et al.: J. Am. Chem. Soc. **117** 2896 (1995)
692. M.D. Musick, C.D. Keating, M.H. Keefe et al.: Chem. Mater. **9** 1499 (1997)
693. Y. Liu, Y. Wang, R.O. Clauss et al.: Chem. Phys. Lett. **298** 315 (1998)
694. R. Parthasarathy, X.-M. Lin, H.A. Jaeger: Phys. Rev. Lett. **87** 186807 (2001)
695. J. Schmelzer, Jr., S.A. Brown, A. Wurl et al.: Phys. Rev. Lett. **88** 226802 (2002)
696. R.C. Doty, H. Yu, C.K. Shih et al.: J. Phys. Chem. **B105** 8291 (2001)
697. T. Ogawa, K. Kobayashi, G. Masuda et al.: Thin Solid Films **393** 374 (2001)

698. R.G. Osifchin, W.J. Mahoney, J.D. Bielefeld et al.: Superlattices Microstruct. **18** 283 (1995)
699. R.G. Osifchin, W.J. Mahoney, J.D. Bielefeld et al.: Superlattices Microstruct. **18** 275 (1995)
700. G. Markovich, C.P. Collier, S.E. Hendricks et al.: Acc. Chem. Res. **32** 415 (1999)
701. G. Medeiros-Ribeiro, D.A.A. Ohlberg, R.S. Williams et al.: Phys. Rev. **B59** 1633 (1999)
702. A. Taleb, F. Silly, A.O. Gusev et al.: Adv. Mater. **12** 633 (2000)
703. T.P. Bigioni, L.E. Harrell, W.G. Cullen et al.: Eur. Phys. J. **D6** 355 (1999)
704. V. Torma, T. Reuter, O. Vidoni et al.: Chemphyschem. **2** 546 (2001)
705. S. Pal, N.S. John, P.J. Thomas et al.: J. Phys. Chem. **B108** 10770 (2004)
706. M. Haruta, N. Yamada, T. Kobayashi et al.: J. Catal. **115** 301 (1989)
707. G.K. Bethke, H.H. Kung: Appl. Catal. **194–195** 43 (2000)
708. M. Valden, X. Lai, D.W. Goodman: Science **281** 1647 (1998)
709. C.P. Vinod, G.U. Kulkarni, C.N.R. Rao: *Surface Chemistry and Catalysis*, ed by A. Carley, P. Davies, G. Hutchings and M. Spencer (Kluwer, Dordecht 2003)
710. R. Narayanan, M.A. El-Sayed: Langmuir **21** 2027 (2005)
711. D.I. Enache, J.K. Edwards, P. Landon: Science **311** 362 (2006)
712. Y. Yin, R.M. Rioux, C.K. Erdonmez et al.: Science **304** 711 (2004)
713. M. Kerker: J. Colloid Interface Sci. **105** 297 (1985)
714. P.K. Jain, K.S. Lee, I.H. El-Sayed et al.: J. Phys. Chem. **B110** 7238 (2006)
715. S.K. Medda, S. De, G. De: J. Mater. Chem. **15** 3278 (2005)
716. G. De, C. N.R. Rao: J. Phys. Chem. **B107** 13597 (2003)
717. G. De, C.N.R. Rao: J. Mater. Chem. **15** 891 (2005)
718. F. Strelow, A. Fojtik, A. Henglein: J. Phys. Chem. **98** 3032 (1994)
719. A. Henglein, D. Meisel: J. Phys. Chem. **B102** 8364 (1998)
720. T. Linnert, P. Mulvaney, A. Henglein: J. Phys. Chem. **97** 679 (1993)
721. A. Henglein: J. Phys. Chem. **B103** 9302 (1999)
722. C.S. Weisbecker, M.V. Merritt, G.M. Whitesides: Langmuir **12** 3763 (1996)
723. S. Berchmans, P.J. Thomas, C.N.R. Rao: J. Phys. Chem. **B106** 4647 (2002)
724. Y. Kim, R.C. Johnson, J.T. Hupp: NanoLetters **1** 165 (2001)
725. C.L. Nehl, H. Liao, J.H. Hafner: NanoLetters **6** 683 (2006)
726. S.C. Salzemann, A. Brioude, M.-P. Pileni: J. Phys. Chem. **B110** 7208 (2006)
727. M.D. Malinsky, K.L. Kelly, G.C. Schatz et al.: J. Am. Chem. Soc. **123** 1471 (2001)
728. A.D. McFarland, R.P. van Duyne: NanoLetters **3** 1057 (2003)
729. G. Laurent, N. Felidj, S.L. Truong: NanoLetters **5** 253 (2005)
730. M. Quinten, U. Kreibig: Surf. Sci. **172** 557 (1986)
731. W. Rechberger, A. Hohenau, A. Leitner et al.: Opt. Commun. **220** 137 (2003)
732. S.A. Maier, M.L. Brongersma, P.G. Kik et al.: Phys. Rev. **B65** 193408 (2002)
733. H. Wang, D.W. Brandl, F. Le et al.: NanoLetters **6** 827 (2006)
734. R. Viswanatha, S. Sapra, S.S. Gupta et al.: J. Phys. Chem. **B108** 6303 (2004)
735. R. Viswanatha, D.D. Sarma: Chem. Eur. J. **12** 180 (2006)
736. R. Seshadri, G.N. Subbanna, V. Vijayakrishnan et al.: J. Phys. Chem. **99** 5639 (1995)
737. V.A. Fonoberov, K.A. Alim, A.A. Balandin et al.: Phys. Rev. **B73** 165317 (2006)
738. Y.S. Wang, P.J. Thomas, P. O'Brien et al.: J. Phys. Chem. **B110** 4099 (2006)

739. E.J.D. Klem, L. Levina, E.H. Sargent: Appl. Phys. Lett. **87** 053101 (2005)
740. G. L Tan, N. Wu, J.G. Zheng et al.: J. Phys. Chem. **B110** 2125 (2006)
741. K. Akamatsu, T. Tsuruoka, H. Nawafune: J. Am. Chem. Soc. **127** 1634 (2005)
742. H. Chander: Mater. Sci. Eng. **R49** 113 (2005)
743. R.D. Schaller, V.I. Klimov: Phys. Rev. Lett. **92** 186601 (2004)
744. R.J. Ellingson, M.C. Beard, J.C. Johnson et al.: NanoLetters **5** 865 (2005)
745. P. Guyot-Sionnest: Nature Mater. **4** 653 (2005)
746. A.J. Nozik: Annu. Rev. Phys. Chem. **52** 193 (2001)
747. A.J. Nozik: Physica (Amsterdam) **14E** 115 (2002)
748. M.J. Bowers II, J.R. McBride, S.J. Rosenthal.: J. Am. Chem. Soc. **127** 15378 (2005)
749. A. Kasuya, R. Sivamohan, Y.A. Barnakovi et al.: Nature Mater. **3** 99 (2004)
750. B. Fisher, J.M. Caruge, D. Zehnder et al.: Phys. Rev. Lett. **94** 087403 (2005)
751. J.Y. Kim, F.E. Osterloh: J. Am. Chem. Soc. **127** 10152 (2005)
752. D. Jurbergs, E. Rogojina, L. Mangolini et al.: Appl. Phys. Lett. **88** 233116 (2006)
753. Y. Kayanuma: Phys. Rev. **B38** 9797 (1988)
754. A.I. Ekimov, A.A. Onushchenko, A.G. Plukhin: JETP **88** 1490 (1985)
755. N.F. Borrelli, D.W. Hall, H.J. Holland et al.: J. Appl. Phys. **61** 5399 (1987)
756. D.M. Mittleman, R.W. Schoenlein, J.J. Shiang et al.: Phys. Rev. **B49** 14435 (1994)
757. A. Henglein: Ber. Bunsenges. Phys. Chem. **99** 903 (1995)
758. A.I. Ekimov, A.A. Onushchenko, A.I. Efros: JETP Lett. **43** 376 (1986)
759. A.I. Ekimov, A.A. Onushchenko, S.K. Shumilov et al.: Sov. Tech. Phys. Lett. **13** 115 (1987)
760. A.I. Ekimov, A.I. Efros, M.G. Ivanov et al.: Solid State Commun. **69** 565 (1989)
761. Z. Lifshitz, M. Yassen, L. Bykov et al.: J. Phys. Chem. **98** 1459 (1994)
762. J.R. Heath, J.J. Shiang: Chem. Soc. Rev. **27** 65 (1998)
763. Y. Wang, N. Herron: Phys. Rev. **B42** 7253 (1990)
764. S. Sapra, D.D. Sarma: Phys. Rev. **B69** 125304 (2004)
765. R. Viswanatha, S. Sapra, T. Saha-Dasgupta et al.: Phys. Rev. **B72** 045333 (2005)
766. R. Viswanatha, S. Sapra, B. Satpati et al.: J. Mater. Chem. **14** 661 (2004)
767. A. Wood, M. Giersig, M. Hilgendirff et al.: Aust. J. Chem. **56** 1051 (2003)
768. E.A. Muelenkamp: J. Phys. Chem. **B102** 5566 (1998)
769. H. Nakamura, W. Kato, M. Uehara et al.: Chem. Mater. **18** 3330 (2006
770. D. Vanmaekelbergh, P. Liljeroth: Chem. Soc. Rev. **34** 299 (2005)
771. M.V. Artemyev, U. Woggon, H. Jaschinski et al.: J. Phys. Chem. **B104** 11617 (1999)
772. C.R. Kagan, C.B. Murray, M.G. Bawendi: Phys. Rev. **B54** 8633 (1996)
773. H. Döllefeld, H. Weller, A. Eychmüller: J. Phys. Chem. **B106** 5604 (2002)
774. H. Döllefeld, H. Weller, A. Eychmüller: NanoLetters **1** 267 (2001)
775. A. Samokhvalov, M. Berfeld, M. Lahav et al.: J. Phys. Chem. **B104** 8632 (2000)
776. M.V. Artemyev, A.I. Bibik, L.I. Gurinovich et al.: Phys. Rev. **B60** 1504 (1999)
777. X. Peng, J. Wickham, A.P. Alivisatos: J. Am. Chem. Soc. **120** 5343 (1998)
778. D. Zhang, K.J. Klabunde, C.M. Sorenson et al.: Phys. Rev. **B58** 14167 (1998)
779. F. Gazeau, J.C. Bacri, F. Gendron et al.: J. Magn. Magn. Mater. **186** 175 (1998)

780. R.V. Upadhyay, D. Srinivas, R.V. Mehta: J. Magn. Magn. Mater. **214** 105 (2000)
781. W.H. Meikeljohn, C.P. Bean: Phys. Rev. **105** 904 (1957)
782. W.H. Meikeljohn, C.P. Bean: Phys. Rev. **102** 1413 (1956)
783. D.A. Dimitrov, G.M. Wysin: Phys. Rev. **B50** 3077 (1994)
784. S. Gangopadhyay, G.C. Hadjipanayis, B. Dale et al.: Phys. Rev. **B45** 9778 (1992)
785. V. Skumryev, S. Stoyanov, Y. Zhang et al.: Nature **423** 19 (2003)
786. *Nanophase Materials*, ed by G.C. Hadjipanayis, R.W. Siegel (Kluwer, Dordrecht 1994)
787. C.M. Sorensen: *Magnetism in Nanoscale Materials in Chemistry* ed by K. J. Klabunde (Wiley-Interscience, New York 2001)
788. C. Liu, B. Zou, A.D. Rondinone: J. Am. Chem. Soc. **122** 6263 (2000)
789. H.T. Zhang, X.H. Chen: Nanotechnology **17** 1384 (2006)
790. C.R. Vestal, Z.J. Zhang: Int. J. Nanotechnol. **1** 240 (2004)
791. O. Masala, R. Seshadri: Chem. Phys. Lett. **402** 160 (2005)
792. Z.X. Tang, C.M. Sorenson, K.J. Klabunde et al.: Phys. Rev. Lett. **67** 3602 (1991)
793. G.U. Kulkarni, K.R. Kannan, T. Arunarkavalli et al.: Phys. Rev. **49** 724 (1994)
794. P.J.v.d. Zaag, V.A.M. Brabers, M.T. Johnson et al.: Phys. Rev. **51** 12009 (1995)
795. C. Liu, Z. J. Zhang: Chem. Mater. **13** 2092 (2001)
796. A.E. Berkowitz, W.J. Shuele, P.J. Flanders: J. Appl. Phys. **39** 1261 (1968)
797. S. Linderoth, P.V. Hendriksen, F. Bødker et al.: J. Appl. Phys. **75** 6583 (1994)
798. F.T. Parker, M.W. Foster, D.T. Margulies et al.: Phys. Rev. **B47** 7885 (1993)
799. T. Okada, H. Sekizawa, F. Ambe et al.: J. Magn. Magn. Mater. **31–34** 105 (1983)
800. D. Lin, A.C. Numes, C.F. Majkrzak et al.: J. Magn. Magn. Mater. **145** 343 (1995)
801. A.H. Morrish, K. Haneda: J. Appl. Phys. **52** 2496 (1981)
802. R.H. Kodama, A.E. Berkowitz, E.J. McNiff, Jr. et al.: Phys. Rev. Lett. **77** 394 (1996)
803. B. Martínez, X. Obradors, Ll. Balcells et al.: Phys. Rev. Lett. **80** 181 (1998)
804. D. Predoi, V. Kuncser, E. Tronc et al.: J. Phys. Condens. Matter **15** 1797 (2003)
805. J. Park, K. An, Y. Hwang et al.: Nat. Mater. **3** 891 (2004)
806. R.H. Kodama, S.A. Makhlouf, A.E. Berkowitz: Phys. Rev. Lett. **79** 1393 (1997)
807. A. Sundaresan, R. Bhargavi, N. Rangarajan et al.: Phys. Rev. **B** (2006)
808. D. Magana, S.C. Perera, A.G. Harter et al.: J. Am. Chem. Soc. **128** 2931 (2006)
809. K. Biswas, K. Sardar, C.N.R. Rao: Appl. Phys. Lett. (2006)
810. I.M.L. Billas, J.A. Becker, A. Chatelain et al.: Phys. Rev. Lett. **71** 4067 (1993)
811. W.D. de Heer: Nanomagnetism in: *Characterization of Nanophase Materials*, ed by Z. L. Wang (Wiley-VCH, Weinheim 2000)
812. I. Billas, J. Becker, A. Chatelain et al.: Science **265** 1682 (1994)
813. A.J. Cox, J.G. Lauderback, L.A. Bloomfield: Phys. Rev. Lett. **71** 923 (1993)
814. B.V. Reddy, S.N. Khanna, B.I. Dunlap: Phys. Rev. Lett. **70** 3323 (1993)
815. K. Yakushiji, F. Ernult, H. Imamura et al.: Nat. Mater. **4** 57 (2005)
816. H. Zheng, J. Li, J.P. Liu et al.: Nature **420** 395 (2002)

817. A.T. Ngo, M.-P. Pileni: J. Phys. Chem. **B105** 53 (2001)
818. A.T. Ngo, M.-P. Pileni: Adv. Mater. **12** 276 (2000)
819. M.-P. Pileni: J. Phys. Chem. **B105** 3358 (2001)
820. P.J. Thomas, M. Rajamathi, P.V. Vanitha et al.: J. Nanosci. Nanotechnol. **5** 565 (2005)
821. C.F. Hoener, K.A. Allan, A.J. Bard et al.: J. Phys. Chem. **96** 3812 (1992)
822. M.Y. Han, W. Huang, C.H. Chew et al.: J. Phys. Chem. **B102** 1884 (1998)
823. Y. Tian, T. Newton, N.A. Kotov et al.: J. Phys. Chem. **100** 8927 (1996)
824. E. Hao, H. Sun, Z. Zhou et al.: Chem. Mater. **11** 3096 (1999)
825. D. Pan, Q. Wang, S. Jiang et al.: Adv. Mater. **17** 176 (2005)
826. M.A. Hines, P. Guyot-Sionnest.: J. Phys. Chem. **100** 468 (1996)
827. M. Danek, K.F. Jensen, C.B. Murray et al.: Chem. Mater. **8** 173 (1996)
828. X. Peng, M.C. Schlamp, A.V. Kadavanich et al.: J. Am. Chem. Soc. **119** 7019 (1997)
829. A. Ishizumi, Y. Kanemitsu: Adv. Mater. **18** 1083 (2006)
830. N. Revaprasadu, M.A. Malik, P. O'Brien et al.: Chem. Commun. 1573 (1999)
831. M.A. Malik, P. O'Brien, N. Revaprasadu: Chem. Mater. **14** 2004 (2002)
832. Y. He, H.-T. Lu, L.-M. Sai et al.: J. Phys. Chem. **B110** 13370 (2006)
833. J.J. Li, Y.A. Wang, W. Guo et al.: J. Am. Chem. Soc. **125** 12567 (2003)
834. I. Mekis, D.V. Talapin, A. Kornowski et al.: J. Phys. Chem. **B107** 7454 (2003)
835. D.V. Talapin, I. Mekis, S. Götzinger et al.: J. Phys. Chem. **B108** 18826 (2004)
836. Y.-W. Cao, U. Banin: Angew. Chem. Int. Ed. **38** 3692 (1999)
837. Y.-W. Cao, U. Banin: J. Am. Chem. Soc. **122** 9692 (2000)
838. R. Xie, U. Kolb, J. Li et al.: J. Am. Chem. Soc. **127** 7480 (2005)
839. H.S. Zhou, H. Sasahara, I. Homma et al.: Chem. Mater. **6** 1534 (1994)
840. A. Hasselbarth, A. Eychmüller, M. Eichberger et al.: J. Phys. Chem. **97** 5333 (1993)
841. Y. Liu, M. Kim, Y. Wang et al.: Langmuir **22** 6341 (2006)
842. R.H. Morriss, L.F. Collins: J. Chem. Phys. **41** 3357 (1964)
843. L. Lu, H. Wang, Y. Zhou et al.: Chem. Commun. 144 (2002)
844. L. Rivas, S. Sanchez-Cortes, J.V. Garcia-Ramos et al.: Langmuir **16** 9722 (2000)
845. Y.W. Cao, R. Jin, C.A. Mirkin: J. Am. Chem. Soc. **123** 7961 (2001)
846. F. Henglein, A. Henglein, P. Mulvaney: Ber. Bunsenges. Phys. Chem. **98** 180 (1994)
847. P. Mulvaney, M. Giersig, A. Henglein: J. Phys. Chem. **97** 7061 (1993)
848. A. Henglein, M. Giersig: J. Phys. Chem. **B104** 5056 (2000)
849. A. Henglein, M. Giersig: J. Phys. Chem. **98** 6931 (1994)
850. A. Henglein, P. Mulvaney, A. Holzwarth et al.: Ber. Bunsenges. Phys. Chem. **96** 754 (1992)
851. A. Henglein, A. Holzwarth, P. Mulvaney: J. Phys. Chem. **96** 8700 (1992)
852. A. Henglein, P. Mulvaney, A. Holzwarth: J. Phys. Chem. **96** 2411 (1992)
853. I. Katsiks, M. Gutierrez, A. Henglein: J. Phys. Chem. **100** 11203 (1996)
854. Y. Heng, P.C. Gibbons, K.F. Kelton et al.: J. Am. Chem. Soc. **123** 9198 (2001)
855. A.F. Lee, C.J. Baddeley, C. Hardacre et al.: J. Phys. Chem. **99** 6096 (1995)
856. J. Rivas, R.D. Sanchez, A. Fondado et al.: J. Appl. Phys. **76** 6564 (1994)
857. C.T. Seip, C.J. O'Connor: Nanostruct. Mater. **12** 183 (1999)
858. J. Lin, W. Zhow, A. Kumbhar et al.: J. Solid. State. Chem. **159** 26 (2001)
859. J. Zhang, M. Post, T. Veres et al.: J. Phys. Chem. **B110** 7122 (2006)

860. N.S. Sobal, M. Hilgendorff, H. Moehwald: NanoLetters **2** 621 (2002)
861. S. Zhou, B. Varughese, B. Eichhorn et al.: Angew. Chem. Int. Ed. **44** 4539 (2005)
862. P.J. Thomas, G.U. Kulkarni, C.N.R. Rao: J. Nanosci. Nanotechnol. **1** 267 (2001)
863. N. Nunomura, T. Teranishi, M. Miyake et al.: J. Magn. Magn. Mater. **177** 947 (1998)
864. W. Stöber, A. Fink, E. Bohn: J. Colloid Interface Sci. **26** 62 (1968)
865. R.K. Iler: Patent no: 2,885,366 (1959)
866. M. Ohmori, E. Matijevic: J. Colloid Interface Sci. **150** 594 (1992)
867. D.M. Thies-weesie, A.P. Philipse: Langmuir **11** 4180 (1995)
868. A.P. Philipse, M.P. van Bruggen, C. Pathmamanoharan: Langmuir **10** 92 (1994)
869. Q. Liu, Z. Xq, J.A. Finch et al.: Chem. Mater. **10** 3938 (1998)
870. M.A. Correa-Duarte, M. Giersig, N.A. Kotov et al.: Langmuir **14** 6430 (1998)
871. A.P. Philipse, A.M. Nechifor, C. Pathmamanoharan: Langmuir **10** 4451 (1994)
872. L.M. Liz-Marzán, M. Giersig, P. Mulvaney: Chem. Commun. 731 (1996)
873. L.M. Liz-Marzán, M. Giersig, P. Mulvaney: Langmuir **12** 4329 (1996)
874. L.M. Liz-Marzán, P. Mulvaney: New J. Chem. 1285 (1998)
875. P. Mulvaney, L.M. Liz-Marzán, M. Giersig et al.: J. Mater. Chem. **10** 1259 (2002)
876. M.A. Correa-Duarte, M. Giersig, L.M. Liz-Marzán: Chem. Phys. Lett. **286** 497 (1998)
877. M. Bruchez, M. Moronne, P. Gin et al.: Science **281** 2013 (1998)
878. D. Gerion, F. Pinaud, S.C. Williams et al.: J. Phys. Chem. **B105** 8861 (2001)
879. D.K. Yi, S.T. Selvan, S.S. Lee et al.: J. Am. Chem. Soc. **127** 4990 (2005)
880. O.G. Tovmachenko, C. Graf, D.J. van den Heuvel: Adv. Mater. **18** 91 (2006)
881. P. Li, J. Moon, A.A. Morrone et al.: Langmuir **15** 4328 (1999)
882. I. Pastoriza-Santos, L.M. Liz-Marzán: Langmuir **15** 948 (1999)
883. V. Eswaranand, T. Pradeep: J. Mater. Chem. **12** 2421 (2002)
884. R.T. Tom, A.S. Nair, N. Singh et al.: Langmuir **19** 3439 (2003)
885. H. Yu, M. Chen, P.M. Rice, S.X. Wang et al.: NanoLetters **5** 379 (2005)
886. O. Masala, R. Seshadri: J. Am. Chem. Soc. **127** 9354 (2005)
887. I.L. Medintz, H.T. Uyeda, E.R. Goldman et al.: Nature Mater. **4** 435 (2005)
888. A.J. Sutherland: Curr. Opn. Sol. Stat. Mater. Sci. **6** 365 (2002)
889. C.M. Niemeyer: Angew. Chem. Int. Ed. **40** 4128 (2001)
890. H. Gu, R. Zheng, X. Zhang et al.: J. Am. Chem. Soc. **126** 5664 (2004)
891. J.P. Zimmer, S-W. Kim, S. Ohnishi et al.: J. Am. Chem. Soc. **128** 2526 (2006)
892. D. Ishii, K. Kinbara, Y. Ishida et al.: Nature **423** 628 (2003)
893. C.Y. Zhang, H. Ma, S.M. Nie et al.: Analyst **125** 1029 (2000)
894. J.O. Winter, T.Y. Liu, B.A. Korgel et al.: Adv. Mater. **13** 1673 (2001)
895. M. Dahan, T. Laurence, F. Pinaud et al.: Opt. Lett. **26** 825 (2001)
896. N.L. Rosi, C.A. Mirkin: Chem. Rev. **105** 1547 (2005)
897. C.-S. Tsai, T.-B. Yu, C.-T. Chen: Chem. Commun. 4273 (2005)
898. R. Elghanian, J.J. Storhoff, R.C. Mucic et al.: Science **277** 1078 (1997)
899. R.C. Jin, G. Wu, Z. Li et al.: J. Am. Chem. Soc. **125** 1643 (2003)
900. R.A. Farrer, F.L. Butterfield, V.W. Chen et al.: NanoLetters **5** 1139 (2005)
901. A.J. Haes, R.P. van Duyne: J. Am. Chem. Soc. **124** 10596 (2002)
902. J.C. Riboh, A.J. Haes, A.D. McFarland et al.: J. Phys. Chem. **B107** 1772 (2003)

903. D.J. Maxwell, J.R. Taylor, S. Nie: J. Am. Chem. Soc. **124** 9606 (2002)
904. E. Oh, M.-Y. Hong, D. Lee et al.: J. Am. Chem. Soc. **127** 3270 (2005)
905. J. Yguerabide, E.E. Yguerabide: Anal. Biochem. **262** 137 (1998)
906. J. Yguerabide, E.E. Yguerabide: Anal. Biochem. **262** 157 (1998)
907. S. Schultz, D.R. Smith, J.J. Mock et al.: Proc. Natl. Acad. Sci. **97** 996 (2000)
908. M. Lippitz, M.A. van Dijk, M. Orrit: NanoLetters **5** 799 (2005)
909. A. Campion, P. Kambhampati: Chem. Soc. Rev. **27** 241 (1998)
910. B. Nikoobakht, M.A. El-Sayed: J. Phys. Chem. **A107** 3372 (2003)
911. Y. Dirix, C. Bastiaansen, W. Caseri et al.: Adv. Mater. **11** 223 (1999)
912. J.Y. Kim, H. Hiramatsu, F.E. Osterloh et al.: J. Am. Chem. Soc. **127** 15556 (2005)
913. S.A. Maier, M.L. Brongersma, P.G. Kik et al.: Adv. Mater. **13** 1501 (2001)
914. J. Lee, A.O. Govorov, N.A. Kotov: Angew. Chem. Int. Ed. **44** 7439 (2005)
915. Q.A. Pankhurst, J. Connolly, S.K. Jones, et al.: J. Phys. D: Appl. Phys. **36** R167 (2003)
916. F. Paul, D. Melville, S. Roath et al.: IEEE Trans. Magn. Mag. **17** 2822 (1981)
917. A. Jordan, R. Scholz, P. Wust et al.: J. Magn. Magn. Mater. **201** 413 (1999)
918. A. Jordan, R. Scholz, K. Maier-Hauff et al.: J. Magn. Magn. Mater. **225** 118 (2001)
919. J. Won, M. Kim, Y.-W. Yi et al.: Science **309** 121 (2005)
920. W.U. Huynh, J.J. Dittmer, A.P. Alivisatos: Science **295** 2425 (2002)
921. T. Hasobe, H. Imahori, P.V. Kamat et al.: J. Am. Chem. Soc. **127** 1216 (2005)
922. I. Robel, V. Subramanian, M. Kuno et al.: J. Am. Chem. Soc. **128** 2385 (2006)
923. S. Guenes, H. Neugebauer, N.S. Sariciftci et al.: Adv. Funct. Mater. **16** 1095 (2006)
924. K.S. Narayan, A.G. Manoj, J. Nanda et al.: Appl. Phys. Lett. **74** 871 (1999)
925. B.O. Dabbousi, M.G. Bawendi, O. Onitsuka et al.: Appl. Phys. Lett. **66** 1316 (1995)
926. S. Chaudhary, M. Ozkan, W.C.W. Chan: Appl. Phys. Lett. **84** 2925 (2004)
927. M.S. Gudiksen, K.N. Maher, L. Ouyang et al.: NanoLetters **5** 2257 (2005)
928. V.L. Colin, M.C. Schlamp, A.P. Alivisatos: Nature **370** 354 (1994)
929. M.C. Schlamp, X. Peng, A.P. Alivisatos: J. Appl. Phys. **82** 5837 (1997)
930. J. Lee, V.C. Sundar, J.R. Heine et al.: Adv. Mater. **12** 1102 (2000)
931. M. Gao, C. Lesser, S. Kirstein et al.: J. Appl. Phys. **87** 2297 (2000)
932. D. Qi, M. Fischbein, M. Drndic, S. Selmic: Appl. Phys. Lett. **86** 093103 (2005)
933. H.-J. Eisler, V.C. Sundar, M.G. Bawendi et al.: Appl. Phys. Lett. **80** 4614 (2002)
934. Y. Chan, J.S. Steckel, P.T. Snee et al.: Appl. Phys. Lett. **86** 073102 (2005)
935. R.D. Piner, J. Zhu, F. Xu et al.: Science **283** 661 (1999)
936. C.A. Mirkin, S. Hong, L. Demers: Chem **2** 37 (2001)
937. D.S. Ginger, H. Zhang, C.A. Mirkin: Angew. Chem. Int. Ed. **43** 30 (2004)
938. R. McKendry, W.T.S. Huck, B. Weeks et al.: NanoLetters **2** 713 (2002)
939. L.M. Demers, C.A. Mirkin: Angew. Chem. Int. Ed. **40** 3069 (2001)
940. X. Liu, L. Fu, S. Hong et al.: Adv. Mater. **14** 231 (2002)
941. M.B. Ali, T. Ondarcuhu, M. Brust et al.: Langmuir **18** 872 (2002)
942. G. Gundiah, N.S. John, P.J. Thomas et al.: Appl. Phys. Lett. **84** 5341 (2004)
943. P.J. Thomas, G.U. Kulkarni, C.N.R. Rao: J. Mater. Chem. **14** 625 (2004)
944. S.-W. Chung, D.S. Ginger, M.W. Morales et al.: Small **1** 64 (2005)
945. S.J. Green, J.I. Stokes, M.J. Hostetler et al.: J. Phys. Chem. **B101** 2663 (1997)

946. D.L. Klein, R. Roth, A.K.L. Lim et al.: Nature **389** 699 (1997)
947. S.H.M. Persson, L. Olofsson, L. Hedberg: Appl. Phys. Lett. **74** 2546 (1999)
948. D.L. Gittins, D. Bethell, D.J. Schiffrin et al.: Nature **408** 67 (2000)
949. V.V. Agrawal, R. Thomas, G.U. Kulkarni et al.: Pramana J. Phys. **65** 769 (2005)
950. B.C. Regan, S. Aloni, K. Jensen et al.: NanoLetters **5** 1730 (2005)
951. N. Krasteva, I. Besnard, B. Guse et al.: NanoLetters **2** 551 (2002)
952. S.M. Briglin, T. Gao, N.S. Lewis: Langmuir **20** 299 (2002)
953. H. Ahn, A. Chandekar, B. Kang et al.: Chem. Mater. **16** 3274 (2004)
954. J.W. Grate, D.A. Nelson, R. Skaggs: Anal. Chem. **75** 1864 (2003)
955. F.P. Zamborini, M.C. Leopold, J.F. Hicks et al.: J. Am. Chem. Soc. **124** 8958 (2002)
956. H. Wohlhen, A.W. Snow: Anal. Chem. **70** 2856 (1998)
957. P.A.S. Jorge, M. Mayeh, R. Benrashid et al.: Appl. Opt. **16** 3760 (2006)
958. J. Ouyang, C.-W. Chu, D. Sieves et al.: Appl. Phys. Lett. **86** 123507 (2005)
959. R.J. Tseng, J. Huang, J. Ouyang et al.: NanoLetters **5** 1077 (2005)
960. M.D. Fischbein, M. Drndic: Appl. Phys. Lett. **86** 193106 (2005)
961. A.O. Orlov, I. Amlani, G.H. Berstein et al.: Science **277** 928 (1997)
962. G.L. Snider, A.O. Orlov, I. Amlani et al.: J. Appl. Phys. **85** 4283 (1999)
963. J.R. Heath, P.J. Kuekes, G.S. Snider et al.: Science **280** 1716 (1998)
964. H. Abelson, D. Allen, D. Coore et al.: Technical Report A. I. Memo 1665, Massachusetts Institute of Technology, Artificial Intelligence Laboratory, August 1999
965. D. Coore, R. Nagpal, R. Weiss: Technical Report A. I. Memo 1614, Massachusetts Institute of Technology, Artificial Intelligence Laboratory, October 1997
966. H. Abelson, D. Allen, D. Coore et al.: Comm. of the Assoc. Comp. Mach. May 2000
967. T. Hatano, M. Stopa, T. Yamaguchi, T. Ota, K. Yamada, S. Tarucha1: Phys. Rev. Lett. **93** 066806 (2004)
968. W.G. van der Wiel, S. De Franceschi, J.M. Elzerman et al.: Rev. Mod. Phys. **75** 1 (2003)
969. D. Loss, D.P. DiVincenzo: Phys. Rev. A. **57** 120 (1998)
970. V. Cerletti, W.A. Coish, O. Gywat, D. Loss: Nanotechnology **16** R27 (2005)

Index

Ag
  nanotriangles, 140
  nanocrystal chains, 72
  nanocrystal rings, 72
  synthesis of,
    nanocrystals, 26, 28, 32–35, 37, 42, 53, 54, 56, 59, 60
    prisms and plates, 62, 63
    rods and anisotropic structures, 60
  two-dimensional arrays of, 75, 78
Ag_2S
  nanocrystal rings, 72
  Two-dimensional arrays, 79
Ag-Pd, nanocrystal synthesis of, 34
Al, Synthesis of, 36
Alcohol reduction, 34
Alkalides, 35
Alkylaluminates, 35
Alloying, 105, 107
AlN, synthesis of, 26, 27, 41, 51
Amorphous computing, 150
Anisotropy constant, 119
Arc discharge, 27
Atomic layer epitaxy, 127
Au
  nanocrystal chains, 69, 71
  nanocrystal rings, 72
  synthesis of, 26, 28, 32–34, 36, 42, 46, 52, 55–63, 67–68
  prisms and cubes, 62
  rods and anisotropic structures, 61
  tetrapods and branched structures, 62–63
  two-dimensional arrays, 76–78
Au-Ag, two-dimensional arrays of, 76
Au-Pt, synthesis of, 33

$BaCO_3$ wires, synthesis of, 65–66
$BaCrO_4$
  prismatic structures, 65
  synthesis of, 65
  two-dimensional arrays of, 80
$BaFe_{12}O_{19}$, synthesis of, 53
Ball milling, 29
Band-gap variation, 22, 100–113
Barium hexaferrite rings, 74
$BaSO_4$ wires, synthesis of, 65
$BaTiO_3$, 48
BHCOAT, 20
Biacore, 108
Binding energy, 6, 15, 66, 97
Blocking temperature, 11, 27, 42, 119, 121, 131
Borohydride reduction, 30, 60
Bremsstrahlung Isochromat spectroscopy, 16
Brust method, 31, 75

Catalysis, 23, 104, 135
CdS
  synthesis of, 26, 27, 29, 38, 42, 44, 48–51, 53, 54, 56–58, 63
    tetrapods, 64
    triangles, 65
    nanotubes, nanorods, 71

nanocrystal rings, 75
properties, 97, 99, 110–113
CdSe
synthesis of, 26, 29, 38-39, 45, 48–51
tetrapods, 63
two-dimensional arrays, 79, 148
CdTe
synthesis of, 26, 45, 48–49, 58, 63
tetrapods, 63
Centrifugal separation, 59
Chaperonin, 81, 133, 138
Charge doping, 132
Charge injection, 67
Charge transport, 66, 84, 103, 143
Charging energy, 17, 93, 101, 146
Charging potential, 101
Chemisorption, 21, 107
Citrate route, 33
Close packing, 6
Cluster beams, 4, 122
Co
synthesis of
nanocrystals, 26, 28, 33–34, 38–39, 43, 46, 48, 53, 57
nanorods, 63
rings, 73, 75
hexagons, 72
CO oxidation, 104
Coercivity, 10, 118, 124
$CoFe_2O_4$,
nanocrystal synthesis of, 45–46, 55
magnetic properties, 11, 119–120, 132
Coherent rotation, 118
Colloidal crystals, 80, 89
CoO
synthesis of, 42, 48
magnetic properties of, 12, 118, 121
two-dimensional arrays, 80
Co_3O_4,
nanocrystal synthesis of, 48
two-dimensional arrays of, 79
properties, 104
CoPt,
synthesis of, 46–47
$CoPt_3$
synthesis of
nanocrystals, 47
rings, 73
properties, 73
two-dimensional arrays of, 78
Core-shell nanocrystals, 106, 117, 125, 140, 143
Ag-Au, 129, 140
Ag-Cd, 129
Ag-Co, 131
Ag-Hg, 129
Ag-In, 129
Ag-Pb, 129
Ag-SiO_2, 132
Au-Ag, 140
Au-Bi, 129
Au-Cd, 129
Au-Fe, 131
Au-Fe_2O_3, 132
Au-Hg, 129
Au-In, 129
Au-Pb, 129
Au-Pd, 105
Au-SiO_2, 132
Au-Tl, 129
CdS-Ag_2S, 126
CdS-CdSe, 126, 130
CdS-PbS, 129
CdSe-CdS, 126–130, 142
CdSe-ZnS, 126, 132, 135, 142
CdSe-ZnSe, 126
Co-Ag, 131
Co-Au, 131
Co-Pt, 131
$CoFe_2O_4$-MnO, 132
Fe-Ag, 131
Fe-Au, 131
HgS-CdS, 129
InAs-CdSe, 128
InAs-GaAs, 128
InAs-InP, 128
InAs-ZnS, 128
InAs-ZnSe, 128
Mn-doped CdS-CdS, 126
Mn-doped CdS-ZnS, 126
Pd_{561}-Ni, 133
$Pd_{561}Ni_{3000}Pd_{150}$, 133
Pt-Cu, 131
CoS, synthesis of, 56
Cosmetics, 135
Coulomb blockade, 18, 101–103, 147
Coulomb staircase, 18, 101–103

Coulombic states, 17
Coupling of electronic states, 22, 114, 115
Cr, synthesis of, 26, 57
Cu
synthesis of
nanocrystal, 26, 29, 31, 34, 46, 50, 53–55, 64, 65
chains, 70
rings and hexagons, 72
optical properties of, 108
CuO, synthesis of, 48, 56
Cu_2O, synthesis of, 48
Cu-Pd, synthesis of, 34, 35
CuS, synthesis of, 53
$Cu_{1-x}S$, synthesis of, 49
Cu_2S
synthesis of nanorods, nanoprisms, nanodisks, 64
two-dimensional arrays, 79

Data storage, 126
Decahedral particle, 7
Defect migration, 129
Delocalization, 115, 125, 126
Density functional calculations, 98
Density of states, 16, 98, 104
Dilute magnetic semiconductors, 122
Dip-pen nanolithography, 143
Dipolar interactions, 115
Directional photoemitters, 117
Discrete dipole approximation, 107
Displays, 118, 126
DNA, 69, 70, 74, 139, 140, 144, 146
Domain, 1, 10, 27, 118
Domain wall motion, 118
Doping nanocrystals, 66, 126
Drude model, 20

Effective mass approximation, 21, 110
Electrical double layer, 3
Electrides, 35
Electrochemical synthesis, 42
Electroluminescent, 142
Electron affinity, 16, 17
Electronic shells, 14, 15, 95, 96
Electronic structure theory, 98
Enzymes, 57, 108
Epitaxial growth, 126, 129

EuS, synthesis of, 51
Exchange anisotropy, 118
Exchange bias, 118, 123
Exchange interactions, 80, 115
Exchange spring magnet, 123
Exciton, 22, 118, 143
Acceptor-like, 110
Donor-like, 110
Excitonic absorption, 22, 106

Fe
synthesis of, 26, 27, 35, 38, 39, 43, 46, 53
FeC, synthesis of, 26, 29
FeCo
synthesis of, 46
two-dimensional arrays, 76
Fe-Fe_2O_3, synthesis of, 26, 27
γ-Fe_2O_3, 118, 121, 124, 144
nanocrystal synthesis of, 45, 48, 55, 56
three dimensional array, 80
Fe_3O_4
synthesis of, 48, 79
two-dimensional arrays, 80
γ-Fe_3N, synthesis of, 29
FeMo, synthesis of, 46
FeP, wires and rods synthesis of, 63
FePd, synthesis of, 46, 47
FePt
synthesis of,
nanocrystals, 46
magnetic properties of, 148
two-dimensional arrays, 123
Five-fold symmetry, 6, 9
Fluorescence, 67, 117, 129, 132, 136, 140
resonance energy transfer, 140
Fluorescent tags, 135
Fluorocarbon, 68
Foam, 29, 56
Free electron model, 20,
Full color emission, 142

GaAs, nanocrystal synthesis of, 51
GaInP, nanocrystal synthesis of, 51
GaN, nanocrystal synthesis of, 40, 51
GaP, nanocrystal synthesis of, 51
Gd, nanocrystal synthesis of, 26, 27, 36
Ge, nanocrystal synthesis of, 26, 37
Geometric shells, 7, 14, 15, 94, 95

Hamaker constant, 82
Hard spheres, 81, 83
Heat Capacity, 95
HgS, nanocrystal synthesis of, 129
HgTe,
  synthesis of, 40, 45
  properties of, 142
Hollow nanocrystals, 105
Hopping conduction, 103
HRTEM, 7, 8
Hyperthermia, 142

Icosahedron, 6-8
In nanowires, synthesis of, 63
InAs, nanoparticle synthesis of, 51, 52
Inert gas condensation, 26, 27
InN, synthesis of, 41, 51
InP
  synthesis of, 41, 51, 53
  two-dimensional arrays, 79
In_3Sn nanowires, synthesis of, 63
Integrating sphere, 115
Interaction energy, 103
Intermediate confinement, 110
Interparticle interactions, 12, 114
Ion sputtering, 26, 27
Ionization energy, 15, 93, 95, 96
Ionization potential, 16, 17
Ir, nanoparticle synthesis of, 42, 46, 54
I-V spectra, 18, 101, 103, 146, 147

Jahn-Teller distortion, 98

Kirkendal effect, 105
Kubo gap, 13

Langmuir-Blodgett, 70, 81
Laser ablation, 26, 28
Laser synthesis, 59
Lattice match, 125
Lattice mismatch, 128, 129
Linear rows, 109, 141
Long-range resonance transfer, 115
Low-dimensional aggregates, 108
Luminescent Semiconductor nanocrystals, 23, 117, 132

Magic nuclearity, 7, 76, 88, 95, 98, 132
Magnetic anisotropy, 10, 11, 118
Magnetic hysteresis, 12, 42, 122
Magnetic properties, 9, 10, 12, 47, 80, 118, 119, 121
Magnetic resonance imaging, 142
Magnetic shells, 123
Magnetization reversal, 10, 11, 118, 149
Magnetocrystalline anisotropy, 118
Maxwell's equations, 19
Melting point, 14, 60, 93, 94
Memory devices, 148
Memory elements, 145
Metal-Insulator-Nonmetal-Insulator-Metal (MINIM) junction, 101
Metal nanocrystals, 20, 22, 23, 30, 42, 60, 67, 69
Metal oxide nanocrystals, 46, 79, 80
Micelles, 78, 126, 54, 55
  Inverted micelles, 54, 55, 62, 65, 76, 126, 132
  Reverse micelle, 79, 84, 108, 131
Microbial synthesis, 56, 57
Microemulsions, 54
Mies theory, 20, 107
MnO
  Nanocrystal synthesis of, 48, 132
  magnetic properties of, 121
$MnFe_2O_4$,
  nanoparticle synthesis of, 45, 54
  magnetic properties, 119, 121
Mn_3O_4, nanoparticle synthesis of, 48
MnS, nanoparticle synthesis of, 49
MoS_2, nanocrystal synthesis of, 26, 29, 53
Molecular dynamics, 6, 17, 27
Molecular springs, 141
Mössbauer spectroscopy, 12, 119
Mott-Hubbard metal-insulator transition, 103
Multi-shell nanocrystals, 128
Multiple exciton generation, 118, 143

Nanocrystal organizations, 69–73, 75, 76, 81–83, 86–88, 103–114
Nanocrystal Synthesis
  Early transistion metals, 33
  Iron oxide cubes and related structures, 64
  Iron Oxides, 48
  Metal Chalcogenides, 38, 45

Metal selenides, 54
Metals, 44
Noble and Magnetic metals, 46
Perovskite oxides, 48
Sonochemical routes, 52
Transition metal dichalcogenides, 38
Nanoelectronics, 135, 145
Nanoparticle over smooth electrode, 141
Nanophosphors, 118
Nanoscale waveguide, 141
Nanoshells, 107, 110
Nanoswitch, 146
Nanothermometers, 141
Neel's Theory, 11, 119
Ni,
nanocrystal synthesis of, 26, 27, 31, 34, 36, 43, 46, 53
Ni-Mn-Fe, synthesis of, 26, 27
NiO
Nanocrystal synthesis of, 42, 48
magnetic properties of, 12, 13, 121
catalysis, 104
NiS nanorods, nanodisks, nanoprisms synthesis of, 64
Ni_3S_4, nanocrystal synthesis of, 49

Oriented attachment of nanocrystals, 71
Ostwald ripening, 29

Particulate technology, 135
PbS
synthesis of, 39, 44, 49, 50, 53, 58
two-dimensional arrays, 79, 86
PbSe
synthesis of, 40, 49, 63
two-dimensional arrays, 79, 80
properties, 103, 114, 118, 143
PbTe, synthesis of, 50
$PbTiO_3$, nanocrystal synthesis of, 48
Pd,
nanocrystal synthesis of, 32, 34, 52–54, 56
two-dimensional arrays, 77, 81
Pd-Pt, nanocrystal synthesis of, 34
Perpendicular magnetization, 124
Perovskite oxides, 48
Phase transfer, 31, 67, 68
Photocurrent, 142, 143
Photoelectron spectroscopy, 3, 15, 16, 84, 96, 97
Photolysis, 42
Plasmon band, 19, 21, 107, 129
Plasmon band dampening, 107
Plasmon resonance band, 21, 106, 110
Polarizing filters, 141
Polyelectrolytes, 68, 86, 102, 142
Polyol method, 34, 46, 63
Pt
nanocrystal ribbons, 70
synthesis of
nanocrystals, 31, 33, 34, 36, 43, 46, 52, 54, 59
cubic, octahedral, cuboctahedral, 63
hexagonal, 78
dendrites, branched structures, 63
two-dimensional arrays of, 80
Pt-Bi, nanocrystal synthesis of, 35
Pyrolysis, 26, 28, 29

Quantum cellular automata, 149
Quantum confinement, 23, 28, 51, 66, 113, 116
Quantum dot emission, 135
Quantum dot labels, 138
Quantum size effects, 14

Radiolysis, 42
Raman spectroscopy, 108
Reactivity, 23, 104
Recording, 148
Rectification, 104
ReO_3
nanocrystal synthesis of, 41
properties, 107, 121
Rings, 72-75
Rh, synthesis of, 34, 42, 46
Ricatti-Bessel functions, 20
Ru, nanocrystal synthesis of, 34, 42, 46

Saturation magnetization, 121
Scaling law, 1, 93, 101
Scanning tunneling spectroscopy (STS), 16, 100, 146
Scherrer formula, 8
Self-assembly, 12, 69, 73, 101, 142, 146, 150

Semiconductor nanocrystals, 17, 21–23, 48, 59, 63, 66, 99, 106, 125
Shape anisotropy, 118
Si, synthesis of, 26–29, 37, 52, 55
Single electron circuits, 101
Single electron devices, 19, 145
Single electron transistors, 19, 145
SiO_2, synthesis of, 26, 29
Size effects, 2, 14, 88, 121
  Heat capacity, 93
  Melting point, 60, 93
  Specific heat, 94, 95
Size-dependent properties, 1, 60, 70, 93
  Magnetic, 9–11, 119, 121
  Optical, 20, 22, 107, 111, 113, 116
Size-induced metal-insulator transition, 95, 107
SmCo, nanocrystal synthesis of, 46, 48
Sn
  nanocubes, 78
SnO, nanocrystal synthesis of, 26, 27
Soft spheres, 81
Sols, 3, 29, 30
Solvated metal atom dispersion (SMAD), 58
Solvothermal synthesis, 38, 42
Spin glass, 121, 122
Spinel ferrites, 119
Spray pyrolysis, 26, 29
Sterically stabilized, 3, 30, 59
Stern-Gerlach Technique, 122
STM, 8, 9, 85, 98
Stokes shift, 117
Strong confinement regime, 110
Successive ion layer adsorption and reaction (SILAR), 127
Superclusters, 88
Superhydride, 33
Superparamagnetism, 12, 42, 118, 136
Supersensitive electrometers, 145
Supported catalyst, 105
Surface enhanced Raman scattering (SERS), 141
Surface plasmon band, 19
Suzuki reaction, 105

Teramac, 151
Textiles, 135
Tight binding approximation, 17, 110–112
TiO_2, nanocrystal synthesis of, 26, 27, 29, 48, 71
Tl, nanocrystal synthesis of, 42
Triethylborohydrides, 32
Tunneling spectroscopy, 99, 103
Twinned particle, 7
Two-dimensional arrays, 74, 76, 80, 83, 103, 123, 133
  with different sizes, 78

Uncompensated spins, 121
Uniaxial anisotropy, 10, 118

Vapor sensors, 147
$VO_{1.7}$, nanocrystal synthesis of, 28

Weak confinement regime, 110
White light, 117, 143

Y_2O_3, synthesis of, 37

ZnO
  synthesis of
    nanocrystals, 37, 42, 45, 48
    disks, rods, cones and other shapes, 65, 68
    tetrapods, 63
  properties, 113, 147
ZnS
  Mn doped ZnS, 126
  nanocrystal ribbons, 70
  synthesis of
    nanocrystals, 39 , 44, 49, 50, 56, 63
    wires and rods, 63
ZnSe, nanocrystal synthesis of, 26, 27, 38, 49, 66
Zsigmondy, 3

Springer Series in
MATERIALS SCIENCE

Editors: R. Hull R. M. Osgood, Jr. J. Parisi H. Warlimont

30 **Process Technology for Semiconductor Lasers**
Crystal Growth and Microprocesses
By K. Iga and S. Kinoshita

31 **Nanostructures and Quantum Effects**
By H. Sakaki and H. Noge

32 **Nitride Semiconductors and Devices**
By H. Morkoç

33 **Supercarbon**
Synthesis, Properties and Applications
Editors: S. Yoshimura and R. P. H. Chang

34 **Computational Materials Design**
Editor: T. Saito

35 **Macromolecular Science and Engineering**
New Aspects
Editor: Y. Tanabe

36 **Ceramics**
Mechanical Properties, Failure Behaviour, Materials Selection
By D. Munz and T. Fett

37 **Technology and Applications of Amorphous Silicon**
Editor: R. A. Street

38 **Fullerene Polymers and Fullerene Polymer Composites**
Editors: P. C. Eklund and A. M. Rao

39 **Semiconducting Silicides**
Editor: V. E. Borisenko

40 **Reference Materials in Analytical Chemistry**
A Guide for Selection and Use
Editor: A. Zschunke

41 **Organic Electronic Materials**
Conjugated Polymers and Low Molecular Weight Organic Solids
Editors: R. Farchioni and G. Grosso

42 **Raman Scattering in Materials Science**
Editors: W. H. Weber and R. Merlin

43 **The Atomistic Nature of Crystal Growth**
By B. Mutaftschiev

44 **Thermodynamic Basis of Crystal Growth**
P–T–X Phase Equilibrium and Non-Stoichiometry
By J. Greenberg

45 **Thermoelectrics**
Basic Principles and New Materials Developments
By G. S. Nolas, J. Sharp, and H. J. Goldsmid

46 **Fundamental Aspects of Silicon Oxidation**
Editor: Y. J. Chabal

47 **Disorder and Order in Strongly Nonstoichiometric Compounds**
Transition Metal Carbides, Nitrides and Oxides
By A. I. Gusev, A. A. Rempel, and A. J. Magerl

48 **The Glass Transition**
Relaxation Dynamics in Liquids and Disordered Materials
By E. Donth

49 **Alkali Halides**
A Handbook of Physical Properties
By D. B. Sirdeshmukh, L. Sirdeshmukh, and K. G. Subhadra

50 **High-Resolution Imaging and Spectrometry of Materials**
Editors: F. Ernst and M. Rühle

51 **Point Defects in Semiconductors and Insulators**
Determination of Atomic and Electronic Structure from Paramagnetic Hyperfine Interactions
By J.-M. Spaeth and H. Overhof

52 **Polymer Films with Embedded Metal Nanoparticles**
By A. Heilmann

Springer Series in
MATERIALS SCIENCE

53 **Nanocrystalline Ceramics**
Synthesis and Structure
By M. Winterer

54 **Electronic Structure and Magnetism of Complex Materials**
Editors: D.J. Singh and D. A. Papaconstantopoulos

55 **Quasicrystals**
An Introduction to Structure, Physical Properties and Applications
Editors: J.-B. Suck, M. Schreiber, and P. Häussler

56 **SiO_2 in Si Microdevices**
By M. Itsumi

57 **Radiation Effects in Advanced Semiconductor Materials and Devices**
By C. Claeys and E. Simoen

58 **Functional Thin Films and Functional Materials**
New Concepts and Technologies
Editor: D. Shi

59 **Dielectric Properties of Porous Media**
By S.O. Gladkov

60 **Organic Photovoltaics**
Concepts and Realization
Editors: C. Brabec, V. Dyakonov, J. Parisi and N. Sariciftci

61 **Fatigue in Ferroelectric Ceramics and Related Issues**
By D.C. Lupascu

62 **Epitaxy**
Physical Principles and Technical Implementation
By M.A. Herman, W. Richter, and H. Sitter

63 **Fundamentals of Ion-Irradiated Polymers**
By D. Fink

64 **Morphology Control of Materials and Nanoparticles**
Advanced Materials Processing and Characterization
Editors: Y. Waseda and A. Muramatsu

65 **Transport Processes in Ion-Irradiated Polymers**
By D. Fink

66 **Multiphased Ceramic Materials**
Processing and Potential
Editors: W.-H. Tuan and J.-K. Guo

67 **Nondestructive Materials Characterization**
With Applications to Aerospace Materials
Editors: N.G.H. Meyendorf, P.B. Nagy, and S.I. Rokhlin

68 **Diffraction Analysis of the Microstructure of Materials**
Editors: E.J. Mittemeijer and P. Scardi

69 **Chemical-Mechanical Planarization of Semiconductor Materials**
Editor: M.R. Oliver

70 **Applications of the Isotopic Effect in Solids**
By V.G. Plekhanov

71 **Dissipative Phenomena in Condensed Matter**
Some Applications
By S. Dattagupta and S. Puri

72 **Predictive Simulation of Semiconductor Processing**
Status and Challenges
Editors: J. Dabrowski and E.R. Weber

73 **SiC Power Materials**
Devices and Applications
Editor: Z.C. Feng

74 **Plastic Deformation in Nanocrystalline Materials**
By M.Yu. Gutkin and I.A. Ovid'ko

75 **Wafer Bonding**
Applications and Technology
Editors: M. Alexe and U. Gösele

76 **Spirally Anisotropic Composites**
By G.E. Freger, V.N. Kestelman, and D.G. Freger

77 **Impurities Confined in Quantum Structures**
By P.O. Holtz and Q.X. Zhao

MIX
Papier aus verantwortungsvollen Quellen
Paper from responsible sources
FSC® C105338

If you have any concerns about our products,
you can contact us on
ProductSafety@springernature.com

In case Publisher is established outside the EU,
the EU authorized representative is:
Springer Nature Customer Service Center GmbH
Europaplatz 3, 69115 Heidelberg, Germany

Printed by Libri Plureos GmbH
in Hamburg, Germany